FORTRAN

FORTRAN

With Emphasis on the CDC Lower 3000 Computer Series

KENNETH P. SEIDEL

GOODYEAR PUBLISHING COMPANY

Pacific Palisades, California

Library of Congress Catalog Card Number: 77-171516

ISBN: 0-87620-321-7

Y-3217-0

Current printing (last number):

10 9 8 7 6 5 4 3 2 1

Printed in the United States of America

CONTENTS

ACKNOWLEDGMENTS

Many people contributed their time and offered their enthusiastic support to me during the writing of this textbook.

Mr. Floyd Dunn of Control Data Corporation cooperated in securing valuable manuals and also was instrumental in securing permissions to quote portions of various technical publications of CDC.

Mr. Hovey G. Reed, director of Institutional Research, Sacramento State College Computer Center, reviewed the manuscript thoroughly, pointing out areas that needed strengthening, clarification, or correction. A special debt of gratitude is owed to him.

Similar thanks are expressed to Mr. H. S. P. Chin of CDC for his thoughtful review.

Other persons who made valuable contributions to this effort were Mr. Robert Langsner, Xerox Data Systems, Mr. Benson C. Stone, senior student at San Fernando Valley State College, and Mr. Doug Maddox of the Cal Poly (Pomona) computer center.

INTRODUCTION

This textbook on Fortran is designed as a general introduction to the American National Standard Fortran computer programming language. Special emphasis is placed on applying the Control Data 3170, 3300, and 3500 computers to problem solving, in both batch processing (MASTER) and time-sharing modes of operation (KRONOS III).

Because of the widespread use of Fortran under the MSOS operating system on similar CDC computers (3100, 3150, 3200), the chapter on deck set-up includes explanations for either the MASTER or MSOS operating environment.

Unless otherwise stated, the Fortran concepts explained here apply to all "lower 3000" computers, using either MASTER or MSOS.

An attempt was made to make the explanations of Fortran as independent of a particular computer system as is possible, without ignoring the fact that specific differences do exist among the various Fortran compilers available in the variety of computer systems in use today. Thus, this text can be used in most computer environments where the Fortran language is supported.

In devising the problems for this text, I have avoided problems of the "make-work" variety, and have instead presented realistic and complete program assignments of gradually increasing complexity from chapter to chapter.

1

FIRST PRINCIPLES

1.1 The First Fortran

In 1957, a revolutionary new concept of how to do scientific programming became available to users of the IBM 704 computer. It was called "Fortran," or more formally, The Fortran Automatic Coding System for the IBM 704.

Fortran, a name made by combining the first and last syllables of "Formula Translation," was a complex computer program and a simple to use, special computer language. The Fortran program was designed to produce programs in the 704's machine language, which would in turn do the things implied in the Fortran statements read by the Fortran Translator.

This development took about three years and involved the efforts of thirteen programmers, designers, and technical writers.

Because of their important roles in this major computing breakthrough, we should mention their names, just as they appeared in the first Fortran Programmer's Reference Manual:

J. W. Backus
R. J. Beeber
S. Best
R. Goldberg
H. L. Herrick
R. A. Hughes

L. B. Mitchell
R. A. Nelson
R. Nutt
D. Sayre
P. B. Sheridan
H. Stern
I. Zeller

Control Data 3150 with removable disk pack mass storage device in foreground.

John Backus has recounted the frenzied days of Fortran's development period in the IBM Data Processing Division *Computing Report* for November

1966. From that article, one can imagine the excitement of implementing a radically new approach to making the computer more productive–an approach that turned out to produce an almost universal computer language that not only makes programs written for one type of computer capable of working on another type of computer, but which also has made the power of the computer available to a wider set of users than just computer programming specialists.

Nowadays, Fortran is available on nearly every computer.

1.2 Source and Object Programs

The Fortran program produced by the original IBM team was called (generically) a compiler, and the term compiler has continued to have the same meaning, not only for the Fortran language, but for other artificial computer languages such as the business-oriented COBOL. A compiler is a generalized computer program that

1. accepts program statements expressed in a particular artificial user-oriented language, and
2. produces a "machine language" equivalent program which can be executed in order to perform the tasks stated by the user's original statements.

The input (step 1) to a compiler is termed the *source program;* the output (step 2) is termed an *object program.*

The whole point of this type of system is that many complex and sophisticated "machine language" programs can be produced by the compiler, automatically, by simply preparing the many different source programs. A considerable improvement in the computer utilization results, since much of the usual error-prone aspect of making computer programs is automated. Furthermore, it is faster and easier to get a working program, by virtue of the fact that humans find it easier to do their program writing tasks in Fortran than in the machine's own language.

The process of compilation is shown schematically in Figure 1.1.

Now, Fortran source programs can be input to the computer for compilation via terminals (such as a Teletype device) connected to a remote computer, in a process called time-sharing. The CDC 3170, 3300, and 3500 computers, for example, possess the capability of preparing a program through the use of the terminal, rather than the use of a keypunch device. Thus, the engineer or student can create, compile, and execute his program in a short time span, rather than having to physically submit a program as a deck of cards and then wait up to several hours for his job to be run and returned to him.

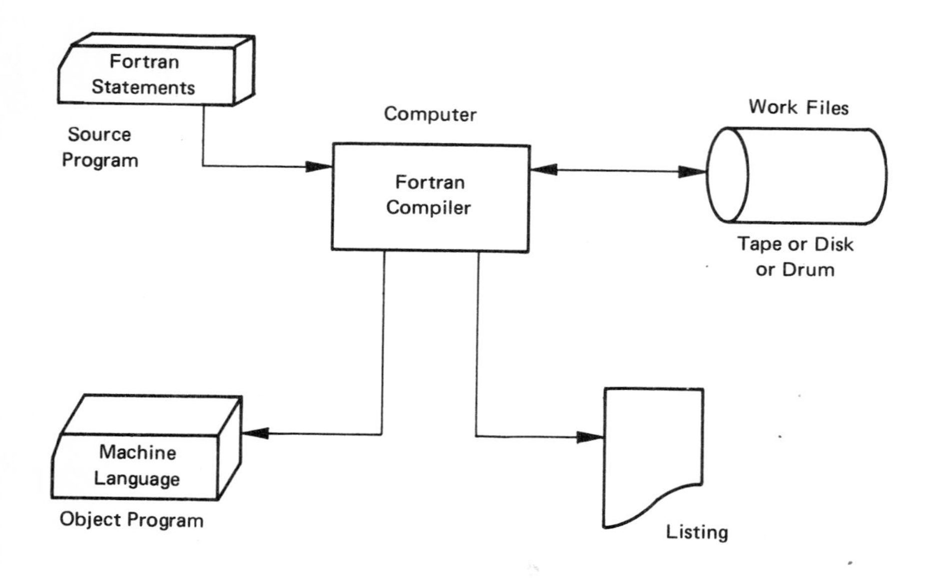

FIGURE 1.1 Fortran compilation process

We will maintain a card-oriented approach to explaining Fortran until Chapter 7, which is devoted to the subject of Fortran in a time-sharing environment.

1.3 Source Statements on Cards

An 80-column punched card, used for Fortran source program input, has 4 fields:

Field	Purpose
Columns 1–5	Statement number, if any
Column 6	Continuation designator
Columns 7–72	Fortran statement
Columns 73–80	Not used by Fortran; typically reserved for deck identification and sequencing.

Whenever a single statement is too long to fit on one card, successive cards may be employed as continuations, provided a continuation number is placed in column 6. Continuation numbers should appear in the order 1 to 9; zero is not permitted as a continuation number.

Figure 1.2 is a replica of a typical Fortran coding form. A programmer may use such a form in order to specify the precise spacing of his statements, since each character position is indicated on the form.

Cards in which the letter C appears in column 1 are considered as comments, and are listed by the compiler without further action.

1.4 Integer and Real Variables

As in mathematics, a symbolic name is given to data "variables." Variables are given names of from one to six characters in length. A name must begin with a letter. The other characters (if any) that make up the name of a variable may be any alphabetic or digit characters. For the purpose of getting started in our study of Fortran, let us consider just two types of variables: real and integer. (This is exactly the limitation in the first Fortran; in the intervening years, other data types have been added; we will consider them later.)

An integer variable cannot assume values possessing a fractional part. Only whole numbers, such as 1, 3, -7, and 0, are permitted as values of an integer variable. Normally, such variables are given names beginning with one of the letters I, J, K, L, M, or N. The maximum value of an integer variable is dependent on the design of the computer in question and of the compiler itself. For the CDC lower 3000 series of computers, the maximum integer value is $2^{47}-1$, which equals 140737488355327.[1]

For more extensive calculations, real variables are used. Real variables are normally named beginning with any letter other than I, J, K, L, M, or N. With this type of data, which is akin to scientific notation, all data (except zero) is represented by a fraction and an exponent. In the CDC lower 3000 series of computers, the minimum non-zero positive number is approximately 10^{-308} —in other words, a number less than 1 in which 307 zero digits follow the decimal point preceding the first non-zero digit! The maximum number is on the order of 10^{308}.

Real numbers are carried in the computer in a form called floating-point. In floating-point on the CDC lower 3000, the fraction has the equivalent of about 11 significant digits. This means that, for a number such as 10^{15}, we would not be able to exactly represent all of its 16 digits. This is analogous to using a slide rule having 11 significant digits of accuracy (as opposed to the generally accepted 3 digits of accuracy on an actual slide rule). It should be noted that possession of 11 significant digits is usually more than enough, considering the accuracy of empirical data used in engineering and scientific data processing. The extra capability is required, however, in order to restrict the propogation of significant

1. Assuming that the compiler reserved two words for storage of a single variable. See option S, section 6.4; see also section 7.1.

CONTROL DATA CORPORATION

FORTRAN CODING FORM

PROGRAM	NAME	
ROUTINE	DATE	PAGE OF

TYPE	STATEMENT NO.	CONT.	FORTRAN STATEMENT: 0 = ZERO, Ø = ALPHA O; 1 = ONE, I = ALPHA I; 2 = TWO, Z = ALPHA Z	SERIAL NUMBER
1	2 3 4 5	6	7 8 9 10 11 12 13 14 15 16 17 18 19 20 21 22 23 24 25 26 27 28 29 30 31 32 33 34 35 36 37 38 39 40 41 42 43 44 45 46 47 48 49 50 51 52 53 54 55 56 57 58 59 60 61 62 63 64 65 66 67 68 69 70 71 72	73 74 75 76 77 78 79 80
1	2 3 4 5	6	7 8 9 10 11 12 13 14 15 16 17 18 19 20 21 22 23 24 25 26 27 28 29 30 31 32 33 34 35 36 37 38 39 40 41 42 43 44 45 46 47 48 49 50 51 52 53 54 55 56 57 58 59 60 61 62 63 64 65 66 67 68 69 70 71 72	73 74 75 76 77 78 79 80

FIGURE 1.2 Fortran coding form

round-off errors as computations are performed based on input data. Although the result of any calculation in floating-point always has an apparent accuracy of 11 significant digits, the actual number of accurate significant digits can be less, and is always limited by the accuracy of the inputs and by the inherent limitations of the numerical method applied.

1.5 Integer and Real Constants

Constants are self-identifying elements of a Fortran program. They do not have names, in contrast to variables.

An integer constant consists of a combination of digits, optionally preceded by a plus or minus sign, and having a numerical value not exceeding $2^{47} - 1$ or not less than -2^{47}.

A real constant consists of a signed or unsigned string of up to 11 decimal digits having either a decimal point or an exponent suffix, or both. An exponent suffix is:

the letter E followed by an integer constant in the range -308 to +307

The following tabulation illustrates various legal constants in the CDC lower 3000 Fortran language.

Constant	Comment
3.14159	Real
0.314159E+1	Real
314159E–05	Real
0	Integer
0.0	Real
0.	Real
+700	Integer

The values of each of the first 3 constants are equivalent, in that by applying the decimal exponents in the exponent suffixes, the second fraction is to be multiplied by 10^1 and the third fraction is to be multiplied by 10^{-5}, yielding 3.14159 in each case.

1.6 Arithmetic Assignment Statement

An assignment statement has the general form

variable = expression

An expression may be a constant, a variable, or an arithmetic operation combining constants and/or variables. The symbols used to form arithmetic operations are

+	addition
-	subtraction
*	multiplication
/	division
**	exponentiation

Using these symbols, we can write assignment statements such as

```
     W = A + B
   TOB = T/B
   SUM = F1 + F2 + F3
     I = J + K
SQUARE = A ** 2
     Q = U
```

(So far, we are limiting the definition of expressions to encompass only facts presented up to this point.)

The meaning of an assignment statement in a program is "perform the calculation indicated by the expression, and assign the resultant value to the variable designated to the left of the = sign."

Because the value of the expression replaces the former value stored in the computer at the place in memory allocated for the left-hand-side variable, the assignment statement is also sometimes called the replacement statement.

It is Fortran's ability to determine the proper sequence of computer program steps to calculate the value of an expression[2] which led to the choice of the name for the language: "*for*mula *tran*slation."

2. Sometimes mistakenly called a formula.

Consider the real expression A + B/C. It should be obvious that the value to be calculated will be dependent on the order of operations. Rules are needed to tell whether the statement means

1. Form the sum A + B, and then divide that result by C, or
2. Form the ratio B/C, and to that result add A.

The rules adopted by the creators of Fortran dictate that the second method is the one actually employed. This rule states that evaluation is conducted from left to right, except when the succeeding operation has a higher "binding power" than the one currently being considered.

The binding power of + and - is lowest in strength; that of * and / is of middle strength; and raising to a power (**) has the highest binding power.

To provide for the possibility of achieving a sequence of operations different from that dictated solely by adhering to the rules of the binding power of operators, the concept of parenthesizing subexpressions was added to the rules for writing expressions. Thus, in order to achieve the equivalent of the mathematical form

$$\frac{A + B}{C}$$

the Fortran programmer writes the expression

```
(A + B)/C
```

Every properly formed expression has an equal number of left and right parentheses, arranged such that, in examining the expression from left to right, there are never more right parentheses in effect than left parentheses.

The following are proper Fortran expressions.

```
((A + B)/C) * DELTA
(A - C)/(B - D)
(-A-1.5 + G7) ** 0.5
```

On the other hand, the following attempts to write Fortran expressions are erroneous.

```
((A / B)
(A- (B- (C + D(4.7))
```

Any constant or variable in an expression must be preceded immediately by one of the following, or be the first constant or variable in the expression:

a left parenthesis
one of the operators +, -, *, /, or **

Any constant or variable in an expression must be followed by one of the following, or be the last constant or variable in the statement:

a right parenthesis
one of the operators +, -, *, /, or **

A unary minus may immediately precede a variable or parenthesized subexpression:

```
- A
- (B/C)
```

Except for the unary minus, each operator +, *, / or ** must have a term or factor both to its left and to its right, in order for an expression to be correctly formed, e.g.,

```
(A + B) / (C - AD)
```

Thus, the statement

```
V = 1.63 * / D
```

is incorrectly formed, and has no meaning.

Unlike conventional mathematical notation, multiplication is not indicated by the juxtaposition of two variable names. In fact, two names side by side could not be distinguished by the Fortran compiler as two distinct names; it would be construed as one name (possibly too long), because of the fact that blanks within a statement are ignored. (Only in connection with a Hollerith constant, to be considered later, is a blank character significant in Fortran.)

The expression Z**Y**W is not valid in Fortran. The calculation must be written as either (Z**Y)**W or Z**(Y**W), whichever is intended.

However, because of the previously stated rules for evaluating expressions, an expression such as

```
A / B / C
```

is valid, and is equivalent to (A/B)/C.

1.7 Division of Integers

When we write I = J/K, we must bear in mind the unusual fact that the computer performs this operation much like a grammar school student. The division operation yields two "answers," one called the quotient, the other called the remainder. Thus, the integer 5 divided by the integer 3 yields

Quotient = 1
Remainder = 2

In our problem I = J/K, we see that our answer I must be an integer. Fortran says use the quotient, and forget the remainder!

On the other hand, if we want an answer 1.6 from dividing 5 by 3, we will use real variables. The following program statements could be used:

```
  A = 3.0
  B = 5.0
EYE = B/A
```

Note that the result is real, and therefore its name does not begin with a letter in the range I–N.

Consider this program sequence:

```
  I = 5
  J = 3
EYE = I/J
```

What result value do you expect in the variable EYE? The correct answer is 1.0, rather than 1.6, because the arithmetic evaluation of the expression is carried out in integer fashion, since that is the "mode" of every operand there. Then, after developing the integer result of the expression, it is converted to the form of a real number for assignment to the variable EYE. Rules for mode change are presented in a later chapter after we learn about other types of data.

1.8 Computer Memory Words

A computer memory (main storage) is a set of many computer *words*. Each word is uniquely accessible by means of an address (a numbered location). Data and computer instructions reside in memory. Program execution is controlled by specifying the required instructions, which involve data or other instruction words.

The word size varies in machines of different types made by different manufacturers. In the CDC lower 3000 series of computers, the word size is 24 bits. A bit is a digit in the binary number system (base 2); a bit value is either 0 or 1. Just as in the decimal number system, the position of a bit affects the value of the number. For example, the binary integer

$$\begin{aligned} 000000000000000000101011 &= \\ 2^5 + 2^3 + 2^1 + 2^0 &= \\ 32 + 8 + 2 + 1 &= \\ 43 \text{ (decimal)} & \end{aligned}$$

A principal advantage of Fortran is that the user need not become familiar with these details of computer instructions and internal number representations.

However, knowledge of computer words is necessary when considering the amount of memory allocated for variables for different types.

2

A BASIC SET OF FORTRAN STATEMENTS

2.1 Statements and Statement Numbers

A Fortran program consists of statements, some executable and some nonexecutable. An executable statement corresponds directly to object program machine instructions that govern the behavior of the computer; nonexecutable instructions supply additional information required to make the process of execution workable.

Statements may be numbered uniquely in order to provide a means of referring to a particular part of the program. Statement numbers appear in card columns 1-5. Not all statements are numbered, however.

We have already learned a great deal about the arithmetic assignment statement. In the following sections of this chapter, we will learn about these statements: IF, GO TO, END, PRINT, FORMAT, and STOP. We will then be able to write a simple Fortran program.

Program execution begins at the first executable statement of a program, and proceeds from there to the next statement, and so on, unless a statement (such as GO TO or IF) specifically directs an alternate path of flow to be taken.

2.2 GO TO Statement

The GO TO statement has three forms. The simplest of these is

GO TO n

where n is the statement number of an executable statement. Execution of this GO TO statement designates the number of the next statement to be executed.

The second form of the GO TO statement is called the "computed" GO TO, and has the general form

GO TO $(n_1, n_2, \ldots n_k), j$

where j is an integer variable and the n_i are statement numbers in a parenthesized list, in which successive statement numbers are separated by a comma. Also, take careful note of the required comma after the right parenthesis.

Execution of a computed GO TO statement depends on the value of j. If $0 < j \leqslant k$, the next executable statement after the computer GO TO statement is numbered n_j. If j lies outside the range 1 to k, the next statement executed is the first executable statement after the computed GO TO statement–in other words, program flow continues sequentially.[3]

The third form of the GO TO statement, called the Assigned GO TO, is regarded by most Fortran experts with distaste. The assigned GO TO statement is

GO TO $m, (n_1, n_2, \ldots n_k)$

where m is a simple integer variable name, and the n_k are statement numbers. The variable m may only be set by means of an ASSIGN statement, not by an arithmetic (replacement) statement and not by an input statement. An ASSIGN statement of the form ASSIGN n TO m must be executed prior to executing the assigned GO TO associated with m. In this latter usage, n must be one of the statement numbers (n_i) in the assigned GO TO. Execution of the assigned GO TO causes control to be passed to the statement whose number was most recently assigned to m.

3. Contrary to this *standard* method of handling the "out of range" case, some compilers have been implemented (inexcusably) to function in a different way. One such compiler is MS FORTRAN under MSOS.

2.3 Arithmetic IF Statement

The arithmetic IF statement provides a three-way "branch," and has the general form

IF (expression) *i*, *j*, *k*

where *i*, *j*, and *k* are statement numbers.

This form of the IF statement tests the value of the expression in parentheses, and directs program flow to executable statement number *i* if the value is negative, to statement number *j* if zero, and to statement number *k* if the value of the expression is positive (positive meaning greater than zero).

Examples of arithmetic IF statements:

```
IF (A-5.0) 10, 17, 84
IF (B)     110, 1, 110
IF (I+K-J) 8, 8, 9
```

From the above examples, we see that any two of the three required statement numbers in the IF statement can be the same, if desired. Of course, it would be nonsense to have all three statement numbers the same in any one IF statement.

2.4 The END Statement

The END statement, which is our first nonexecutable statement, must be the last statement of any Fortran source program. The statement consists only of the word END; it tells the compiler that the complete source program has been read in.

2.5 The Formatted PRINT Statement

The PRINT statement provides a means of producing printed output. Although the PRINT statement is not a part of the USASI Fortran standard, most compilers nevertheless accept the statement as valid Fortran since it is a holdover from the original version of Fortran. In particular, the formatted PRINT statement is available to users of CDC Fortran, IBM OS/360 Fortran, and Univac 1108 Fortran, as well as most others.

The formatted PRINT statement has the general form

PRINT *f*, list

where *f* is the statement number of a FORMAT statement appearing elsewhere in the program, and list is a series of one or more variable names (separated by commas if there is more than one variable in the list).

The values of the listed variables are printed in accordance with specifications contained in the designated FORMAT statement.

Examples of formatted PRINT statements:

```
PRINT 909, A, B, C
PRINT    6, I
PRINT   77, I, J, BVALUE, DELTA, RATIO
```

2.6 Beginner's Facts about FORMAT Statements

A FORMAT statement has the general form

FORMAT (specification)

and in addition must have a statement number. A FORMAT statement may appear anywhere before the END statement.

The specifications enclosed in parentheses provide information about the manner in which variables are to be prepared for output. We will be concerned with the use of the FORMAT statement for purposes of input in a later discussion.

For each variable in a list in an output statement (such as PRINT), there is a specification to determine the output format. Multiple specifications are separated by commas.

The specification

I7

tells the computer to output an integer in seven print positions. The general form of the I specification is

I*w*

where *w* is an unsigned integer field "width." The user chooses a value of *w* sufficiently large to provide space for the maximum or minimum value, including one character position for the sign (positive values are represented by a blank;

negative values are represented by a minus sign) immediately prior to the most significant digit. Leading zeros are suppressed (represented by blanks). Output is such that the units position is the right-most character of the *w* characters that are output.

Real numbers are output under control of either the F or E specification.

The general form of the F specification is

F*w.d*

where *w* is the field width and *d* determines the number of decimal places.

Output of a real number under control of the F format specification converts the associated value to decimal form (rounded in the dth decimal place, in CDC lower 3000 Fortran).[4] Choice of w and d must be such that

$$w \geqslant d + 2,$$

allowing for d digits in the fraction in addition to room for the actual decimal point and the leading sign representation. If the real value being output is positive, the sign is output as a blank character; otherwise a minus sign appears. The *w* value must also allow for the maximum number of digits left of the decimal point.

Failure to supply a field width sufficient for representation of a value of a variable results in *w* asterisks being output, in both I and F format specifications.

Before considering the E format specification, let us consider some examples using I and F specifications, by using a program fragment as an example.

```
      A = 1.5E+0
      B = -2.667
      C = 0.89E+5
      D = 4.4E-2
      I = 31
      J = 747
      PRINT 747, A, B, C, D, I, J
 747  FORMAT (F5.2,F8.2,F7.0,F10.6,I3,I5)
      .
      .
      .
```

FIGURE 2.1 Program fragment illustrating FORMAT

4. In many other non-CDC systems, this rounding is not performed.

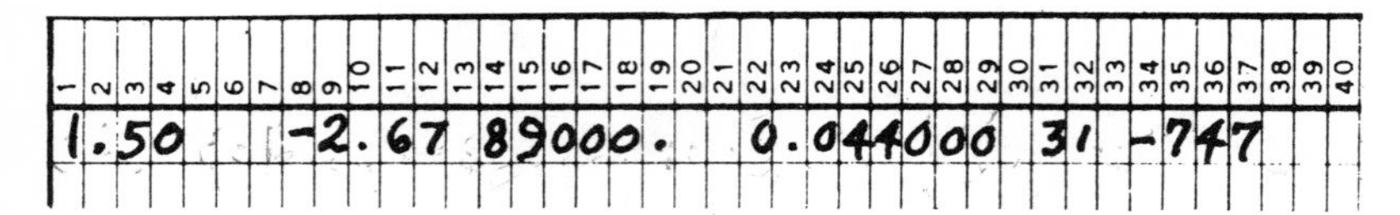

FIGURE 2.2 Output print positions for program fragment in Figure 2.1

The first character of a line of characters developed by a FORMAT statement is not printed by the PRINT statement. Instead, the character is interpreted as a line advancement control character, according to the table in Figure 2.3. The maximum number of characters which may be printed per line is 136.[5]

First Character	Printer Action
Blank	Advance one line before printing; results in "single spaced" output.
0	Advance two lines before printing; results in "double spaced" output.
1	Advance to the top line of the next page before printing (this is called "eject").
+	No line advance; causes "over-printing."

FIGURE 2.3 Print line control characters

Thus, of the 38 characters developed in Figure 2.2 from execution of the program fragment in Figure 2.1, only the last 37 are actually printed, and single spacing results due to the blank first character.

5. Maximum for most non-CDC computers is 132.

The E format specification provides the capability of outputting a real value in the form of a real constant consisting of a fraction and an exponent. The general form of the E specification is

E*w*.*d*

The associated value has *d* digits to the right of the decimal place. The sign of the value appears just to the left of the decimal place as - for negative numbers or blank for positive or zero values. Field width *w* must be sufficient to contain the fraction portion, sign, decimal point, and 4 character exponent suffix; i.e., $w \geqslant d + 6$. The *w* value chosen is usually larger to permit spaces between successive output numbers.

Since the exponent portion can be so large in the CDC lower 3000 computers, there are three possible forms of the exponent suffix:

E-ee when $-99 \leqslant$ exponent < 0
Eeee when $0 \leqslant$ exponent $\leqslant 308$
-eee when $-308 \leqslant$ exponent $\leqslant -100$

Example of E format specification:
Assume that X has the value 167.23.

```
     PRINT 1, X
   1 FORMAT (E 14.5)
```

Result: .16723E003
preceded by four blank characters.

2.7 The STOP Statement

The STOP statement terminates execution of a program. The STOP statement has the general form

STOP *n*

where *n* is optional. If *n* is present, it is up to 5 digits, none of which exceeds 7.

Execution of a STOP statement causes a print-out of *n* and then the program containing the STOP statement relinquishes control to the operating system.

2.8 Sample Program Run

Figure 2.5 is an actual compile-and-go run of a program key punched from the coding sheet shown in Fig. 2.4.[6] In the figure, the first three control cards are

$JOB
$SCHED
$FTNU

as shown (except the $ does not print), followed by the source program, followed by

FINIS
$OBJ

followed by data and the end-of-job card. Refer to Chapter 6 for details of the control cards. The ones shown in this figure may be used by the student, except his name should be substituted in the $JOB card in lieu of SEIDEL. The following is a statement of the problem for Figures 2.4 and 2.5.

Write a program to compute the roots of a quadratic equation, if it has real roots. For the equation

$$y = ax^2 + bx + c$$

adopt the notation A, B, and C for the coefficients (which should be variables, in the Fortran sense). Use assignment statements to provide particular values of A, B, and C such that $b^2 - 4ac$ is not negative. Compute the roots X1 and X2 from use of the quadratic formulae

$$x_1 = \frac{-b + \sqrt{b^2 - 4ac}}{2a}$$

$$x_2 = \frac{-b - \sqrt{b^2 - 4ac}}{2a}$$

6. Note that the programmer writes the letter O with a slash through it to differentiate from the digit zero.

FORTRAN Coding Form

```
C     CALCULATE RØØTS ØF A QUADRATIC EQUATIØN.
C     IF RØØTS ARE NØT REAL (DISCRIMINANT NEGATIVE),
C     STØP IMMEDIATELY.
C     1.  SET UP CØEFFICIENT VALUES, THEN CØMPUTE DISCRIMINANT.
      A = 1.5
      B = 9.8
      C = 0.63
      DISCRM = B ** 2 - 4.0 * A*C
C     2.  TEST FØR NEGATIVE DISCRIMINANT.
      IF (DISCRM) 16, 14, 14
14    SURD = DISCRM ** 0.5
      D = 2.0*A
      X1 = (-B+SURD)/D
      X2 = -(B+SURD)/D
      PRINT 15, A, B, C, X1, X2
15    FØRMAT(F9.3, F9.3, F9.3, E18.9, E18.9)
16    STØP 16
      END
```

FIGURE 2.4 Sample program

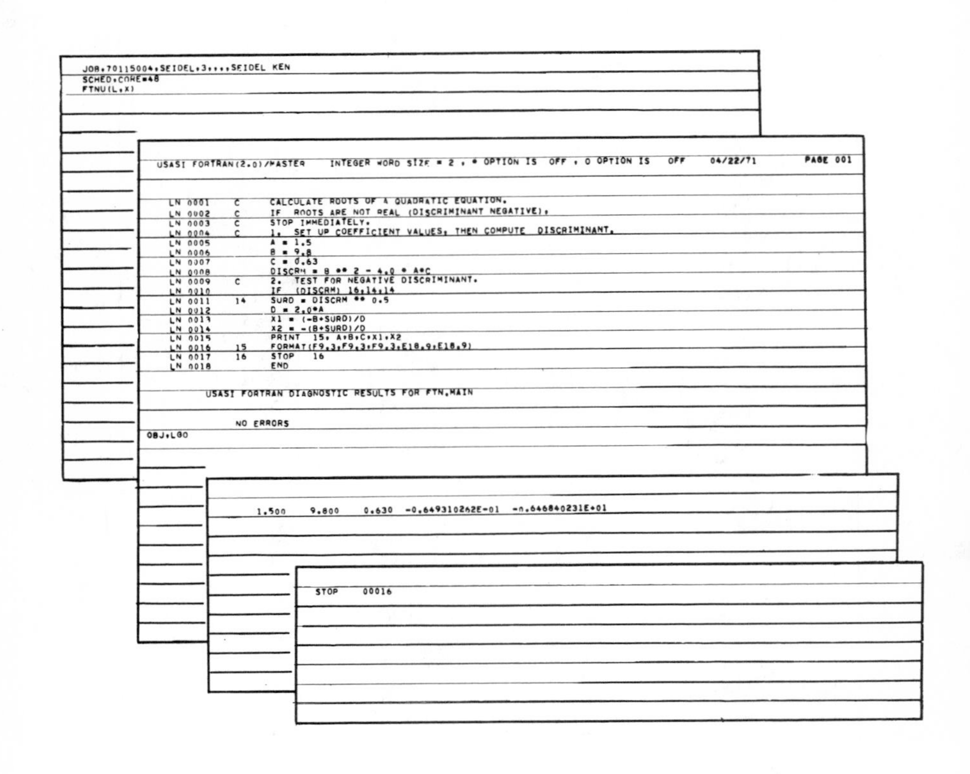

```
JOB,70115004,SEIDEL,3,,,,SEIDEL KEN
SCHED,CORE=48
FTNU(L,X)
```

```
USASI FORTRAN(2.0)/MASTER    INTEGER WORD SIZE = 2 , * OPTION IS  OFF , O OPTION IS   OFF    04/22/71        PAGE 001

  LN 0001    C      CALCULATE ROOTS OF A QUADRATIC EQUATION.
  LN 0002    C      IF  ROOTS ARE NOT REAL (DISCRIMINANT NEGATIVE),
  LN 0003    C      STOP IMMEDIATELY.
  LN 0004    C      1.  SET UP COEFFICIENT VALUES, THEN COMPUTE  DISCRIMINANT.
  LN 0005           A = 1.5
  LN 0006           B = 9.8
  LN 0007           C = 0.63
  LN 0008           DISCRM = B ** 2 - 4.0 * A*C
  LN 0009    C      2.  TEST FOR NEGATIVE DISCRIMINANT.
  LN 0010           IF  (DISCRM) 16,14,14
  LN 0011    14     SURD = DISCRM ** 0.5
  LN 0012           D = 2.0*A
  LN 0013           X1 = (-B+SURD)/D
  LN 0014           X2 = -(B+SURD)/D
  LN 0015           PRINT  15, A,B,C,X1,X2
  LN 0016    15     FORMAT(F9.3,F9.3,F9.3,E18.9,E18.9)
  LN 0017    16     STOP   16
  LN 0018           END

        USASI FORTRAN DIAGNOSTIC RESULTS FOR FTN.MAIN

              NO ERRORS
OBJ,LGO
```

```
     1.500     9.800     0.630  -0.649310262E-01  -0.646840231E+01
```

```
STOP     00016
```

FIGURE 2.5 Sample program run

2.9 Flowcharting

Programmers and analysts find it useful and convenient to use a flowchart in the planning stage of program specifications. A flowchart is generally considered to be an essential item of documentation. The basic idea of flowcharting is to depict the relationships between the logical steps of a program.

In flowcharting, there are two basic conventions:

1. A functionally separate step is depicted in a rectangular box, from which only one arrow emerges, signifying that the next program step is always as represented in the next box.
2. A decision (or branch) is represented by a rhombus ("diamond") shape, from which multiple arrows emerge, representing different paths taken as the result of examining data for different characteristics. In Fortran, an IF statement always corresponds to a diamond-shaped branch of a flowchart.

Figure 2.6 is a flowchart representing the problem for Figures 2.4 and 2.5. Study Figure 2.6 to see how the problem has been broken down into a series of simple but related steps, which advance toward the desired objective in a systematic manner.

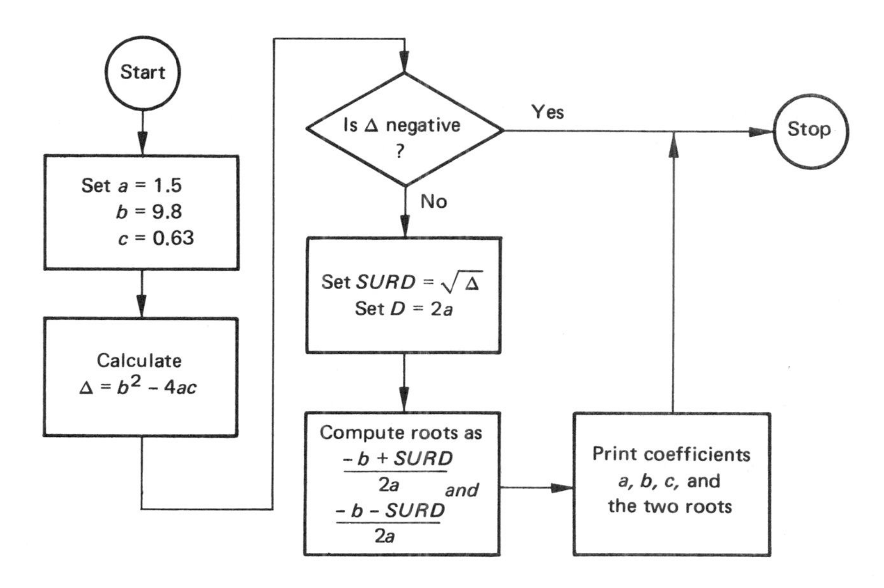

FIGURE 2.6 Flowchart of real roots program

Exercises

1. What statement is the last in every source program to be compiled?
2. Write a simple program that will prove how integer division behaves when either the numerator or denominator (but not both) is negative. Describe the results in a generalized manner, i.e., if $i < 0$ and $j > 0$, what value is computed for $k = i/j$ when $|i| \geqslant |j|$. Also consider the case $i > 0$ and $j < 0$. Is $k = 0$ whenever $|i| < |j|$?
3. The Fibonacci series of integers (0,1,1,2,3,5,8,. . .) is defined by:

$$
\begin{aligned}
k_1 &= 0 \\
k_2 &= 1 \\
k_i &= k_{i-1} + k_{i-2} \text{ for } i \geqslant 3.
\end{aligned}
$$

Write a simple program to generate the integer series one term at a time, and to evaluate the ratio (real) of newest term to its immediate predecessor. Print each term and ratio. Then repeat the same process, using the newest term and its predecessor as the two predecessors from which the next new term will be derived. This series is of interest because the ratio converges on

$$\frac{\sqrt{5+1}}{2},$$

known as the Golden Ratio. The series is of considerable mathematical interest: see *Scientific American,* March 1969, p. 116.

3

ARRAYS AND MORE DATA TYPES

3.1 Array Declarations (DIMENSION Statement)

Lists, arrays, vectors, matrices, and tables are common elements of mathematical interest. Such related collections of data variables can be defined in Fortran by use of the DIMENSION statement.

The general form of the non-executable DIMENSION statement is

$$\text{DIMENSION } \nu_1 \text{ (dimension}_1\text{)}, \nu_2 \text{ (dimension}_2\text{)} \ldots$$

where ν_i are variable names and dimension$_i$ are one of the following for 1-dimensional arrays, 2-dimensional arrays, or 3-dimensional arrays, respectively:

(integer)
(integer, integer)
(integer, integer, integer)

Whenever a variable is defined in a DIMENSION statement, a set of computer memory locations is reserved for the array of variables. Particular members of the array are referred to by the use of subscripts.

Example:

DIMENSION X(5), I(3)

reserves

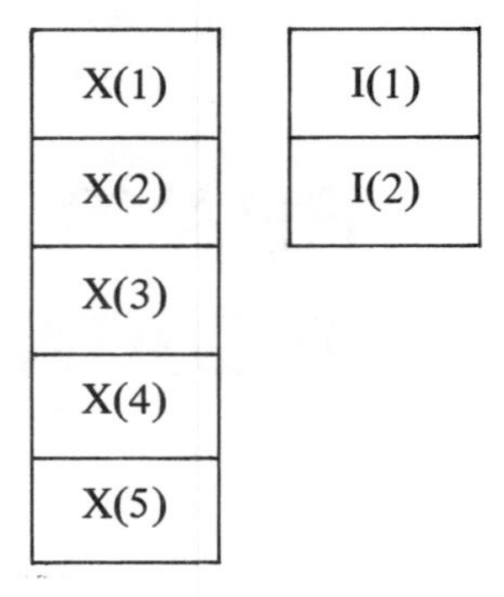

As can be seen from the above, there are 5 X variables, referenced by the subscripts 1, 2, 3, 4, and 5. The number given in parentheses in a DIMENSION statement is the maximum subscript. Subscripts are always integers, and the minimum subscript value is 1.

Consider the statement

DIMENSION A (4,3)

which reserves 12 real variables. Schematically, this can be shown as indicated in Figure 3.1.

A(1,1)	A(1,2)	A(1,3)
A(2,1)	A(2,2)	A(2,3)
A(3,1)	A(3,2)	A(3,3)
A(4,1)	A(4,2)	A(4,3)

FIGURE 3.1 A two-dimensional array

Allocation of an array in computer storage is such that successive positions in memory correspond to varying the left-most subscript first. For a 2-dimensional array, this means that the first column is stored consecutively followed by the second column, and so forth, where A(i,1) designates the first column (for all values $1 \leqslant i \leqslant 4$).

Whenever a variable name may appear in a statement, a subscripted array reference may appear instead. Thus, we may write

```
      XXX = X(2,2) + Y(1,1)
  BETA(3) = BETA(1) + BETA(2)
    A(I,J) = Z(I) * W(J)
```

In the last example above, we employed integer variables I and J as subscripts. The general form of a subscript is

$c * v + d$

where c and d are integer constants, and v is a simple (nonsubscripted) integer variable. Therefore the following statements containing subscripted data references are valid.

```
PRINT 11, A(I), A(2*I), A(I-2)
K(3*L+2) = (M(1)+M(L)+5)/3
```

If a subscript expression (such as 3*L+2) exceeds the maximum value of the associated dimension, as declared in an earlier DIMENSION statement, a run-time error occurs; the results are unpredictable. This is also true if a subscript expression is less than 1, when evaluated during execution.

Every DIMENSION statement must appear with other specification statements prior to the first executable statement of the program.

3.2 Double Precision Variables and Constants

A double precision variable may be thought of as a real variable possessing a larger number of significant digits. The DOUBLE PRECISION statement permits the user to define variables of this type, explicitly. If a DOUBLE PRECISION variable is an array, the array dimensions may be defined in the same statement or in a DIMENSION statement. Any name may be assigned to a double precision variable; since this type of data is declared explicitly, the implicit classification of variables by type depending on the first letter of the name is not applicable.

Examples of DOUBLE PRECISION statements:

```
DOUBLE PRECISION DBL, I(7), B2(2,4)
DOUBLE PRECISION SET (3,10,2)
```

Corresponding to this type of data is a special form of constant, written like a real constant except that the exponent suffix is required and the letter E is replaced by D.

Example:

```
DOUBLE PRECISION PI
PI = 3.1415926536D0
```

3.3 Complex Variables and Constants

Fortran supports complex variables. A complex variable consists of a pair of real numbers representing the real and imaginary parts of a complex number having the general mathematical form

$$a + i \cdot b$$

where i^2 is defined as -1.

In order to define a complex variable in Fortran, the COMPLEX type declaration statement is used. The COMPLEX statement may specify the dimensions of a complex variable array, if desired, instead of using a DIMENSION statement.

Example:

```
COMPLEX Z, ZBAR, TABLE(8), JAY
```

A complex constant is written in the form

$$(a,b)$$

where a and b are real constants. Mathematical operations performed with real variables and constants obey the mathematical rules for complex numbers.

Examples:

```
COMPLEX T, Z, ZBAR, JAY, QUANTY
        T = Z + ZBAR
QUANTY = ((1.5,7.2) + Z)/JAY
```

3.4 Logical Variables and Constants

A logical variable is defined as one having either of two possible states (or values): True or False.

Logical variables are defined by an explicit type statement having the word LOGICAL followed by the desired variable names and their array dimensions, if the variable is an array.

Example:

LOGICAL L, TEST(7), PSEUDO

The logical constants are written as

.TRUE. or
.FALSE.

A logical variable may be assigned a value by use of a logical assignment statement.

3.5 Logical Assignment Statement

A logical assignment statement has the general form

$lv = le$

where *lv* is a logical variable and *le* is a logical expression.

A logical expression is a combination of logical constants and variables, using the operators

.NOT.
.AND.
.OR.

The general form of a logical expression may be defined as

$L_1 \; op_1 \; L_2 \; op_2 \; L_3 \ldots$

where the L_i are logical expressions and op_i is either .AND. or .OR.

A logical expression may be parenthesized, and may be preceded by the logical negation operator .NOT.

At its simplest, a logical expression is either a logical variable or a relational expression.

A relational expression is of the form

arithmetic expression rel arithmetic expression

where rel may be

.NE. signifying not =
.EQ. signifying =
.LE. signifying < or =
.GE. signifying > or =
.LT. signifying <
.GT. signifying >

Figure 3.2 illustrates the use of logical variables, relations, and expressions.

```
       LOGICAL    A, B, C, D, E(2)
       DIMENSION X(3)
    53 A = X(1) .GT. X(2)
    54 B = X(1) .NE. X(2)
    55 C = A .OR. B
    56 D = .NOT. (C .OR. E(2)) .AND. B
```

FIGURE 3.2 Logical examples

The basic rules for determining the state of a logical expression are summarized in Figure 3.3.

Value of Variable		Value of Expressions		
A	B	A or B	A and B	Not A
True	True	True	True	False
True	False	True	False	False
False	True	True	False	True
False	False	False	False	True

FIGURE 3.3 Basic truth table

Assume that the variables of Figure 3.2 have the following initial values:

E(1) = .TRUE.
E(2) = .FALSE.
X(2) = 1.5
X(1) = -100E+0;

then statement number 53 is the evaluation of the relational question

$-100 > 1.5?$

and the answer is "no," so the result is A = .FALSE.

Statement number 54 evaluates the relational question

$-100 \neq 1.5?$

and the answer is yes, so B = .TRUE.

Continuing with statement number 55,

C = A .OR. B

is equivalent to C = .FALSE. .OR. .TRUE.; by applying the truth table of Figure 3.3, we find that

C = .TRUE.

Now consider statement 56. First we evaluate

(C .OR. E(2))

which is .TRUE. since C is true. Therefore

NOT (C .OR. E(2))

is .FALSE.

The final step is to replace everything prior to .AND. B by .FALSE.; regardless of the value of B, the overall expression is .FALSE. in our example.

In evaluating logical expressions, the binding power of operations is (1) NOT (2) AND (3) OR, in order of most to least binding. Of course, parentheses can be used to alter the order of evaluation, as we have seen.

3.6 Logical IF Statement

The logical IF statement has the general form

IF (*le*) executable statement

where *le* is a logical expression evaluatable as .TRUE. or .FALSE. If the value of *le* is .TRUE. then the single statement which follows the parenthesized logical expression is executed. The executable statement component of a logical IF must not itself be a logical IF or DO statement.

Examples (I is an integer variable; LV1 and LV2 are type LOGICAL variables):

```
35 IF (I .EQ. 3) GO TO 747
36 IF (LV1 .AND. LV2) K = K + 1
```

In statement 35, control is transferred to the statement numbered 747 if the value of I is 3, *i.e.,* if the relation I .EQ. 3 is true.

In statement 36, if *both* logical variables LV1 and LV2 are .TRUE. then the value of K is incremented by one; otherwise, the statement does nothing.

3.7 REAL and INTEGER Type Declarations

Real variables may be given a name beginning with any letter, including one of the letters I through N, if a REAL type declaration precedes the first use of the variable name. Dimensions may also be declared in the statement, if desired.

Example:

```
REAL I, J(7), DELTA
```

Note: in the above, DELTA need not have been declared, since it would be classified as a real variable automatically, in the absence of any other type of declaration.

Similarly, integer variables may be given a name beginning with a letter outside the range I through N by an explicit INTEGER type declaration statement.

Example:

```
INTEGER YEAR, VALUES (15)
```

3.8 Mixed Mode Arithmetic Expressions and Assignments

In the CDC lower 3000 USASI[7] Fortran, arithmetic expressions may contain a mixture of any types of data, while the formal USASI standard permits mixing only real with complex and real with double-precision.

The CDC implementation adheres to the following rules:

1. Order of dominance of the element types within an expression, from highest to lowest, is:

Complex (C)	Real (R)
Double Precision (D)	Integer (I)

2. The type of an evaluation arithmetic expression is the mode of the dominant element type.

7. United States of America Standards Institute, now called American National Standards Institute (ANSI).

3. The type relationship and mode of the result for A**B are:

A**B		B Type			
		I	R	D	C
A Type	I	I	R*	D*	C*
	R	R	R	D	C*
	D	D	D	D	C*
	C	C	C*	C*	C*

*Non-USASI usage.

Examples:

1. Given: real A, B; integer I, J. The type of expression A*B–I+J is real because the dominant element type is real. The expression is evaluated in the following ordered steps, where ——► means "is computed and stored as the value of."

A*B ——► R_1	real
Convert I to	real
R_1 -I ——► R_2	real
Convert J to	real
R_2 +J ——► R_3	real

2. Given: C, complex; D, double-precision; R, real; I, integer. The expression C*D+R/I–R is evaluated according to the following steps.

Convert D to	complex
C*D ——► R_1	complex
Convert I to	real
R/I ——► R_2	real
Convert R_2 to	complex
R_2 +R_1 ——► R_3	complex
Convert R to	complex
R_3 -R ——► R_4	complex

3. If the expression in example 2 is changed to C*D+(R/I-R), the evaluation procedure is:

Convert D to	complex
C*D ⟶ R_1	complex
Convert I to	real
R/I ⟶ R_2	real
R_2-R ⟶ R_3	real
Convert R_3 to	complex
R_1+R_3 ⟶ R_4	complex

The above discussion pertains only to the method of evaluating an expression, determining the type of the single result. If the type of this result is different from that of the variable on the left hand side of an arithmetic assignment statement, conversion is undertaken in order to store the result as the new value of the variable, according to Figure 3.4.

Type of v	Type of e	Assignment
Integer	Integer	Assign value
Integer	Real	Truncate fractional part, convert to integer form, assign value
Integer	Double precision	Truncate fractional part, convert to integer form, assign value
Integer	Complex	Truncate fractional part of real part, convert to integer form, assign value
Real	Integer	Convert to real form, assign value
Real	Real	Assign value
Real	Double precision	Assign as much precision as possible
Real	Complex	Assign real part of value
Double precision	Integer	Convert to double-precision real form, assign value
Double precision	Real	Extend real to fill double precision
Double precision	Double precision	Assign value
Double precision	Complex	Extend real part of value to fill double precision
Complex	Integer	Convert to real form, assign value to real part, assign zero imaginary
Complex	Real	Assign value to real part, assign zero imaginary
Complex	Double precision	Assign as much precision as possible to real part, assign zero imaginary
Complex	Complex	Assign value

FIGURE 3.4 Type conversions (v=e) in mixed mode assignments

3.9 The DATA Statement

The DATA statement provides a means of initializing variables to particular values at compilation time. (All other methods consume execution time to assign initial data values.)

The general form of the DATA statement is

$$\text{DATA } k_1/d_1/, k_2/d_2/, \ldots, k_n/d_n$$

where each k_i is a list of variables, array elements, or array names implying the entire array, and each d_i is a list of constants, any one of which may be preceded by a repeat factor, j*, where j is an integer constant.

If a list contains more than one pair of variable and value lists, successive pairs are preceded by a comma.

Examples of DATA statements:

1. DIMENSION ARRAY (5), MATRIX (2)
   DATA ARRAY/5*0.01/, MATRIX /2*0/ Each of the 5 members of ARRAY is set to 0.01 and the 2 members of MATRIX are set to an integer value of 0.
2. REAL X(3,2)
   DATA X/1.5, 2*3.0, -6.0, 0.0, -1./, T/1.0E-10/ The program will begin with initial values in the array X as follows:

   X(1,1) = 1.5
   X(2,1) = 3.0
   X(3,1) = 3.0
   X(1,2) = -6.0
   X(2,2) = 0.0
   X(3,2) = -1.0

   Also, variable T is initialized to 10^{-10}.

3.10 The EQUIVALENCE Statement

The EQUIVALENCE statement enables the programmer to specify that two variables or array identifiers occupy the same storage locations. The general form is:

$$\text{EQUIVALENCE } (a_1), (a_2), \ldots, (a_n)$$

Each a_i is a list of at least two simple or subscripted variable identifiers. All elements of a given list refer to the same storage location. A simple variable can be equivalenced to an array element. Equivalent arrays need not be the same length.

The number of subscripts with an array name must correspond in number to the dimensionality of the array declarator or may be a single subscript value. The array element successor function below defines a relation by which multidimension subscripts are converted to a single subscript value.

Given: DIMENSION A(L, M, N)
to find: A(i, j, k) = A(n)

$$n = i + (j-1)*L + (k-1)*L*M$$

For example, if DIMENSION A(4,3,5) were declared for real array A, then A(46) = A(2,3,4) by substituting $i = 2, j = 3, k = 4, l = 4, m = 3, n = 5$.

Rules for EQUIVALENCE statements:

1. All EQUIVALENCE statements must appear within the specification statement group prior to the first executable statement of the program unit.
2. Only one element of an equivalence set can appear in a COMMON statement. (Refer to chapter 8 for a discussion of COMMON.)
3. An identifier used as a dummy argument cannot be used in an EQUIVALENCE statement.
4. An EQUIVALENCE statement can extend the length of common but it cannot change the origin of a common block or rearrange the order of storage reservation in a common block.
5. An identifier can appear more than once in an EQUIVALENCE statement.
6. An identifier that appears in a COMMON statement and also appears in an EQUIVALENCE statement becomes the base identifier for the equivalence set. When no identifier in an equivalence set belongs to common, the identifier with the lowest storage address becomes the base identifier. All other elements in the equivalence set refer to the base identifier.

Examples (assuming that the two-words-per-integer option is in effect):

1. Align elements of two arrays:

```
DIMENSION A(10,10), I(100)
EQUIVALENCE (A(1), I(1))
```

   In this example, the 100 elements of the integer array I occupy the same storage locations as the 100 elements of real array A.

2. Extend length of common block:

```
COMMON A
DIMENSION A(5), B(5)
EQUIVALENCE (A(3),B(1))
```

The resulting storage allocation is:

Storage Word Addresses	Array A	Array B
P, P+1	A(1)	
P+2, P+3	A(2)	
P+4, P+5	A(3)	B(1)
P+6, P+7	A(4)	B(2)
P+8, P+9	A(5)	B(3)
P+10, P+11		B(4)
P+12, P+13		B(5)

3. Multiple use of identifiers:

```
EQUIVALENCE (A,B), (C,D), (E,F), (A,F), (B,D)
```

The following statement accomplishes the same storage allocation:

```
EQUIVALENCE (A,B,C,D,E,F)
```

Exercises

1. Programming the solution to Wallis' equation by the bracketing method:

$$f(x) = x^3 - 2x - 5 = 0$$

Notation:

- X is desired root; starting value 3
- DX is an abscissa adjustment, initially 0.1
- Y is the value of Wallis' quadratic at X
- XZ is the adjacent abscissa (does not mean X times Z)
- Z is the value of Wallis' quadratic at XZ

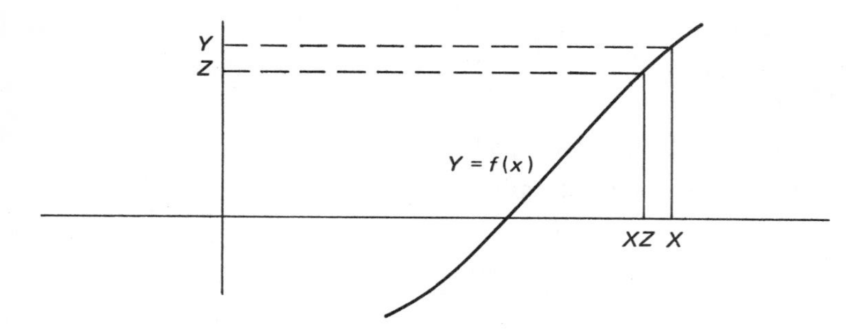

Synopsis of the Method. The starting value $X = 3$ obviously is too high an estimate, while $X = 2$ is too low. We resolve to begin at $X = 3$ and evaluate $X^3 - 2X - 5$ at a point $XZ = X - DX$. We determine whether Z and Y have opposite signs, in which case X and XZ represent abscissa values that are to the right and left of the root. If the signs do not differ, we decrease X by DX and evaluate $f(x)$ at a new XZ, until we have "bracketed" the root, at which time we reduce DX to a smaller step size and continue to advance on the root from the last X. The iterative approach is concluded when $DX < 10^{-11}$.

Statement of the Problem. Develop and run a Fortran program to solve for the roots of Wallis' equation. Show intermediate results in order to illustrate the convergence of this iterative approach. Exercise care in your choice of output format specifications, so that enough decimal places are shown to demonstrate the idea of convergence.

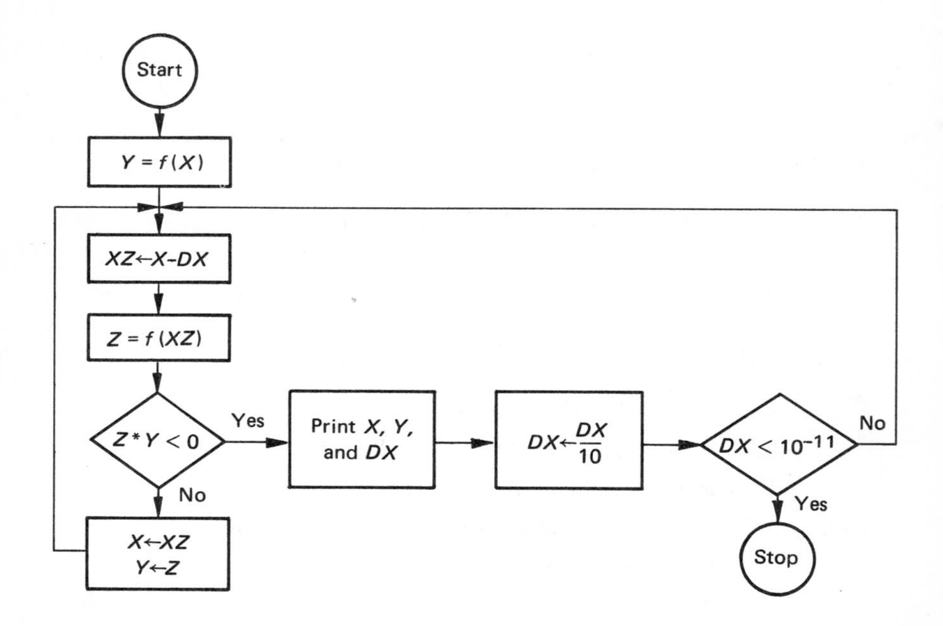

Flowchart of the program

2. No book on Fortran would be complete without a problem based on the *Newton-Raphson method.* The Newton-Raphson method specifies that if an approximate value is available as a root of an equation

$$f(x) = 0$$

then the next (and better) approximation of the root may be calculated by

$$x_{n+1} = x_n - \frac{f(x_n)}{f'(x_n)}$$

where $f'(x)$ represents the first derivative of the function $f(x)$.
Consider the quadratic equation

$$x^2 - A = 0$$

whose root is well known to be $\sqrt{A}$.
Applying the Newton-Raphson method to

$$f(x) = x^2 - A$$
$$f'(x) = 2x$$

then we have

$$x_{n+1} = x_n - \frac{(x_n{}^2 - A)}{2x_n} = \frac{1}{2}(x_n + \frac{A}{x_n})$$

Repeated application of the iteration formula above yields sucessively more accurate approximations of the square root of A. This iteration can be done in Fortran DOUBLE PRECISION, in order to develop a very accurate approximation of the square root. The only thing that must be done is choose a starting value, $x_i \doteq \sqrt{A}$. Crude starting values can be $X_1 = A/2$ for $A > 1$, or even $X_1 = A$. From this your program can derive, in successive steps, X_2, X_3, and so forth.

Program and test this iteration technique in DOUBLE PRECISION, and study the output from each stage of the calculation. The program should terminate the calculations automatically when

$$\left| 1 - \frac{(x_i)^2}{A} \right| < 10^{-15}$$

4

FORMATTED INPUT—OUTPUT

4.1 Formatted READ and WRITE Statements

FORMAT statements serve as "directions" that determine the acquisition of incoming data or control the production of outgoing data. Besides the I, E, and F format specifications (which we have considered only in terms of output, so far), there are several more specifications, including ones oriented toward complex, double precision, and logical data types. Let's first consider how to accomplish input and output in a general way.

Acquisition of input data under format control is accomplished by the READ statement, whose general form is

READ (*i*, *f*) list

where *i* is a unit number (integer constant or variable), and *f* is a FORMAT statement number.

Formatted output is produced by the WRITE statement having the general form

WRITE (*i*, *f*) list

By proper choice of unit number, data may be read from or written to external storage media other than the standard system assignments. Unless otherwise altered by the CDC user, the following fixed unit numbers apply:

Unit	Standard Assignment
60	Normal input (INP)
61	Normal printed output (OUT)
62	Normal punched card output (PUN)

A very good programming technique is to specify the unit number by an integer variable name, which is set to the proper constant value at the beginning of a program. This makes a program less dependent on the unit number standards peculiar to a given installation or computer system, and simplifies the task of converting a Fortran program to a different system at a later time. In general, adherence to such good practice is a requirement in the "real world," where the life of a program often exceeds that of a particular computer at a computer installation.

On input, the variables of a list are garnered from the input unit under control of I, E, F, or other specifications in a FORMAT statement. There is a one-to-one correspondence between the variables in a list and format specifications.

When used for input, the I*w* specification indicates the field width from which an integer is to be obtained. Numbers input under the I format must be justified right in the field of *w* characters. When a signed number is to be read, its sign should immediately precede the first digit.

Example:

Input Card

	∧ ∧ 2	∧ ∧ 2 ∧	∧ ∧	-2	unused
columns	1-3	4-7	8-9	10-11	12-80

Note: ∧ represents one blank.

```
7  READ (INPUT, 1) I, J, K, L
1  FORMAT (I3, I4, 2I2)
```

Execution of statement 7 results in values as follows:

I = 2

J = 20

K = 0
L = -2

In this example, we see that blanks are treated as zero digits. Also, note that the third format specification, I2, was preceded by a repeat factor, so that the third and fourth variables in the list were each obtained from fields of width 2 characters.

When used for input, the F*w.d* specification indicates the input field width, principally. When there is an explicit decimal point in the input data field, the *d* component of F*w.d* is ignored, since the input value is self-defining. Otherwise, the input is treated as an integer, to be multiplied by 10^{-d} in order to obtain the value of the associated real variable in the list.

Example:

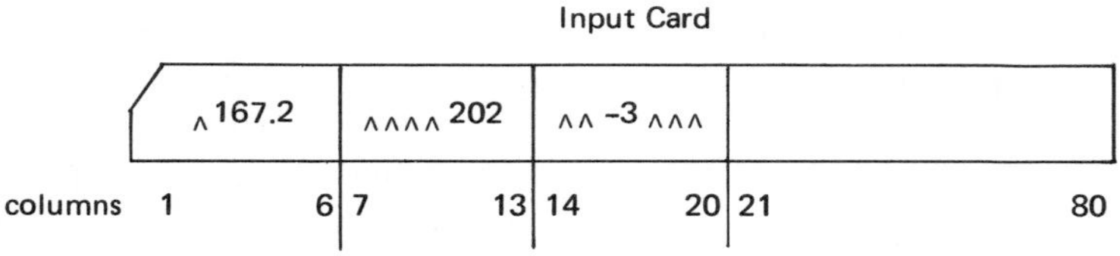

Input Card

∧167.2	∧∧∧∧202	∧∧-3∧∧∧	
columns 1 — 6	7 — 13	14 — 20	21 — 80

Note: ∧ represents one blank.

```
8  READ (INPUT,2) A,B,C
2  FORMAT (F6.0,2F7.2)
```

Execution of statement 8 results in values as follows:

A = 167.2
B = 2.02
C = -30.0

On input, the E*w.d* specification for real data controls a left-to-right scan of the next *w* characters of the input medium. The input field may be in the form of a real constant containing a decimal point, in which case the *d* specification is ignored. The exponent suffix may be omitted.

When the input field does not have an actual decimal point, the scale factor 10^{-d} is applied, regardless of the presence of an exponent suffix.

Thus, if the input specification is E7.7 (which is legal, because the rule $w \geqslant d + 6$ is an output restriction only), and the input field contains the 7 characters

3267+05

the resultant value is

$$3267 \times 10^5 \times 10^{-7} = 32.67$$

The following tabulation of data cards, read under format E10.4 and printed under format E20.8, illustrates how the E format rules are applied, and the resultant values that are stored in a real variable. Note that, when there is no decimal point in the input, the scale factor 10^{-4} is applied to the given integer portion (mantissa), but when there is an actual decimal point in the data the value is as stated in the input data, and the decimal specification (.4) in E10.4 is ignored.

Data, Card Columns 1–10	Output (E20.8)
1234567890	.12345679E006
34567890	.34567890E004
125E0	.12500000E–01
125.E0	.12500000E003
125.	.12500000E003
5.6+03	.56000000E004

In the last example tabulated above, note that there is an exponent not preceded by E. This is permissible only when the data is justified right within the full field width. There should never be embedded blanks in the midst of a field.

Violation of these rules may result in abnormal program termination.

4.2 Non-Data Format Specifications

Hollerith information (named after the inventor of the punched card) consists of textual information. It is used to produce labeling identification, such as titles, heading of tabular output, comments, and carriage control.

The general form of a Hollerith specification is

nH

followed immediately by exactly n characters of textual information. Blanks after nH *are* counted as part of the Hollerith text.

On output, the n characters are output. On input, n characters are read from the input unit, replacing those of the format. (This latter usage is somewhat rare, in practice.)

In either input or output, no list variable is associated with Hollerith information. As an example, we often see statements such as the following, to print "report headings."

```
      WRITE (NØUT, 99)
   99 FORMAT (7H1TØTALS)
```

In the above illustration, the output is the 7 characters

```
1TOTALS
```

but the 1 is used to eject the page, and TOTALS is printed at the left of the top of a new page.

Space may be "skipped over" by use of the specification

*n*X

where *n* represents the number of spaces to be skipped.

For example, to center[8] the heading "TOTALS", the following alternative Format statement could be used.

```
   99 FORMAT (1H1, 65X, 6HTOTALS)
```

4.3 The New-Record Specification

A slash, signaling the end of a record, may appear anywhere in the format specifications. It may be separated from other format elements by commas; consecutive slashes may appear. During output, the slash is used to start new records, cards, or lines. During input, / specifies the beginning of the next card or record.

Example:

```
      WRITE (K2,10)
   10 FORMAT (20X, 7HHEADING///6X,5HINPUT,19X,6HOUTPUT)
```

8. In CDC computers, the printer has 136 characters/line.

4.6 Format Control

Execution of a formatted READ or formatted WRITE statement initiates format control. Each action of format control depends on information jointly provided by the next element of the input/output list, if one exists, and the next field descriptor obtained from the format specification. If there is an input/output list, at least one field descriptor other than *n*H or *n*X must exist.

When a READ statement is executed under format control, one record is read when the format control is initiated, and thereafter additional records are read only as the format specification demands.

When a WRITE statement is executed under format control, a record is written each time the format specification demands that a new record be started. Termination of format control causes writing of the current record.

Except for the effects of repeat counts, the format specification is interpreted from left to right.

For each specification (other than *n*H or *n*X or /) interpreted in a format, there is a corresponding element specified by the input/output list, except that a complex element requires the interpretation of two F, E, or G specifications. To each H or X basic descriptor there is no corresponding element specified by the input/output list, and the format control communicates information directly with the record. Whenever a slash is encountered, the format specification demands that a new record start or the preceding record terminate. During a READ operation, any unprocessed characters of the current record will be skipped at the time of termination of format control or when a slash is encountered.

Whenever the format control encounters a format specification, it determines if there is a corresponding element specified by the input/output list. If there is such an element, it transmits appropriately converted information between the element and the record, and proceeds. If there is no corresponding element, the format control terminates.

If the format control proceeds to the last right parenthesis of the format specification, a test is made to determine if another list element is specified. If not, control terminates. However, if another list element is specified, the format control demands a new record and control reverts to the group repeat specification terminated by the last preceding right parenthesis, or if none exists, then to the first left parenthesis of the format specification.

Examples of format control interaction with an input/output list:

1. List longer than format specifications

```
      DIMENSION NO(4)
      WRITE (6,10) NP,TMAX,PMAX,DRPD,G,C,P1,T1,H1,S1
   10 FORMAT (4I10/(4F10.2))
```

The result is four output records (lines):

```
       101        98       121        97
   1337.28    540.68      -.47     53.70
     71.20    123.00   -823.23       .00
      -.25
```

2. List shorter than format specifications

```
      WRITE (6,10) N1,N2,I,J,A,B
   10 FORMAT (4I10/(4F10.2))
```

Result:

```
       100       200        10        20
       .23    100.00
```

4.7 Alphanumeric Data Handling

The A*w* specification permits the transmission of alphanumeric information in 6-bit character[9] form into the associated variable of the list.

The reader is cautioned that this feature of Fortran differs among computers of different manufacturers. For example, in the GE 600 series or Univac 1100 series computers, where a real variable occupies one 36-bit word, 6 characters can be stored in such a variable.

In the CDC lower 3000 computers, real variables have the capacity of storing up to 8 characters; integer variables store 8 characters (unless the short form option is in effect, in which case each integer variable can store only up to 4 characters). For complex or double precision variables, which occupy 4 words, up to 16 characters may be stored.

If *w* exceeds the number of characters for a variable, only the rightmost characters of the input are stored in the associated variable. If *w* is less than the maximum for a variable, the input characters are stored left justified in the variable, with the remaining character positions being set to the blank character representation.

On A*w* output, when *w* is larger than required, the characters in the variable are output right justified, preceded by blank spaces; if *w* is less than is required, only the leftmost *w* characters from the variable are output.

The R*w* specification deals with alphanumeric information stored in a right justified fashion. Furthermore, zeros are used to fill the remaining character positions, when the field width is less than the storage capacity of the associated variable.

9. In some computers, such as the IBM System 360, one character occupies 8 bits.

4.8 Other Format Specifications

Without discussing them in detail, let us mention the existence of two other types of FORMAT Specifications:

1. G*w.d*–Magnitude-dependent output of real numbers, with or without the exponent part, depending on the value in relation to field width.
2. O*w*–Integers in octal form (base 8 number system)

Exercises

Design and code a FORTRAN program to:

1. Read the total amount (in $ & ¢) purchased by a person, and the amount tendered in payment for that purchase. Assume that neither amount exceeds $5.00. Print these values as read; include dollar signs.
2. Compute and print the total amount of change due, or a message if no change is due or too little was tendered.
3. Print the break-down of the change into dollar bills, quarters, dimes, nickels and pennies. The change is always to be the least number of coins possible, e.g.:

```
AMOUNT PURCHASED    $ 1.08
AMOUNT TENDERED     $ 2.00
----------------------------------------------------
AMOUNT OF CHANGE    $  .92  CONSISTING OF . . .

                              m  QUARTERS
                              n  DIMES
                              k  NICKELS
                              j  PENNIES
```

4. The program must be capable of working on any number of cases, one at a time, each case being represented by one data input card. If any of the quantities *m, n, k,* or *j* is zero, it would be nice to omit the associated line of output.
5. Design a comprehensive series of test cases and determine if the program is always exactly correct.

5

DO LOOPS, ARRAY MANIPULATIONS, AND FUNCTIONS

5.1 The DO and CONTINUE Statements

A convenient method of dealing with subscripted arrays of data is provided by the DO statement, which has the general form

DO $n\ m = i, j, k$

where

n is a statement number,
m is an integer variable,

and

i, j, k are integer constants or integer variables (non-subscripted).

The effect of a DO statement is to set up a range of statements embracing all the statements subsequent to the DO statement, up to and including the one having statement number n.

Execution of the DO range commences with the DO-variable (m) initialized to the value i. Execution then proceeds through the DO range; upon completion of execution of the statement numbered n, the following action takes place.

1. Increase m by k. (If k and its preceding comma were omitted, the value $k = 1$ is used in this step.)
2. The value i is compared to j; if $i > j$, the next statement executed is the first executable one following statement number n. Otherwise ($i \leqslant j$), the entire DO range is re-executed, using the new value of i. This repeated execution range is called a "loop."

Figure 5.1 illustrates use of the DO statement to form the sum of 12 real numbers stored in an array named SET; the sum is retained in a real variable called SUM.

```
      DIMENSION SET (12)
      SUM = 0.0
      DO 9 I = 1,10
    9 SUM = SUM + SET (I)
      PRINT 10, SUM
   10 FORMAT (E20.7)
```

FIGURE 5.1 Simple DO loop

DO loops may be nested, as often occurs in connection with manipulating multi-dimensional arrays. Figure 5.2 illustrates a program sequence to initialize a 2-dimensional array by reading one real value per card.

```
      DIMENSION Q (4,5)
      DO 111 ICOLUM = 1,5
      DO 111 IROW = 1,4
  111 READ (INUNIT, 110) Q(IROW,ICOLUM)
  110 FORMAT (E22.11)
```

FIGURE 5.2 Nested DO loops

Upon reflection, we might wish for a more economical method of inputting data for an array. The availability of DO-implied lists is one possible solution, discussed later in this chapter.

Now, however, let us consider a DO loop in which alternate paths may exist. For example, consider the task of counting how many positive numbers are stored in an array C. Solution of this problem requires that a test (IF statement) be in the loop, so that the last executable statement in the loop, which performs the tallying function, is bypassed in some cases. In this case, we need to use a CONTINUE statement at the end of the range, so that the looping process can continue (see Figure 5.3).

```
   DIMENSION C (52)
   NTALLY = 0
30 DO 33 LOC = 1,52,1
31 IF (C(LOC)) 33,33,32
32 NTALLY = NTALLY + 1
33 CONTINUE
34 next statement . . .
```

FIGURE 5.3 CONTINUE statement illustration

Execution of the program fragment shown in Figure 5.3 is illustrated in flowchart form by Figure 5.4.

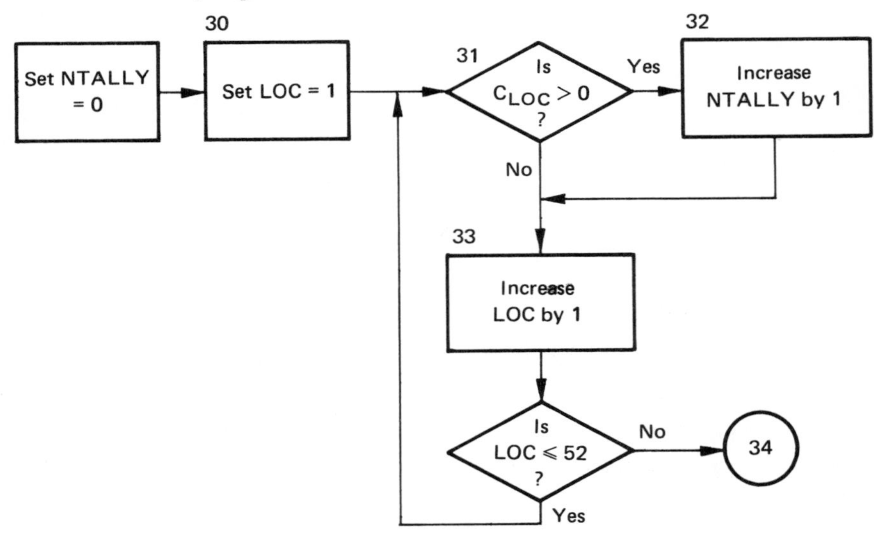

FIGURE 5.4 Flowchart for figure 5.3

Upon completion of a DO loop, the value of the index variable (m) is not necessarily well defined, and should not be referenced.

In order for the DO loop to perform properly, i, j, and k must all be positive at the time of execution of the loop.

The statement designated as the last in a DO loop must not be any of the following: GO TO, FORMAT, DO, IF, RETURN, STOP, or PAUSE.

When two or more DO loops have the same terminal statement, the process of index variable incrementation for the innermost DO loop is executed first. However, if two or more DO loops terminate at the same logical point, and a transfer (say, by an IF or GO TO) is made to the terminal statement for the outer loop, then the inner DO loop must have a separate terminal statement.

As a matter of good programming practice, nested DO loops should have separate CONTINUE termini, as in the following example.

```
      DO    100 I = 1,7
              .
              .
              .
      DO     80 K = 2,4
              .
              .
              .
   80 CONTINUE
              .
              .
              .
  100 CONTINUE
```

DO-implied Lists

DO-implied lists express the variation of subscripted array references in lists of an input or output statement in a manner prescribed explicitly by the user. These are best described by examples. The statement

```
READ (INUNIT, 110)((Q (IROW, ICOLUM), IROW = 1,4), ICOLUM = 1,5)
```

is equivalent to the middle three statements of Figure 5.2. The statement PRINT 32, (A(I), I = 2,17,3) will print the values of A(2), A(5), A(8), A(11), A(14) and A(17). DO-implied subscript variation is such that the innermost DO-type index "runs through" one complete set of values, after setting the values of the outer index(es). This facility is of use whenever partial or selective array input-output is desired.

5.2 Library Function References

We have seen previously that an arithmetic expression consists of arithmetic operators, variable references, and constants.

The Fortran language also permits use of a function reference wherever a variable or constant might appear in an arithmetic expression.

A function reference consists of the function name followed by a parenthesized list of arguments. For example, one may specify "the square root of x" by writing SQRT (X).

There are many built-in function names known to Fortran. Figure 5.5 contains a tabulation of the function names, their definitions, and a characterization of argument and result types.

Examples of function references:

```
IF ( ABS(X-Y) - 1.0E-7 ) 11,11,5
ZBAR = CONJG(Z)
ZBAR = ( REAL(Z), -AIMAG(Z) )
THETA = ATAN (8.6219)
COSINE = COS (ANGLE/57.29578)
```

It should be pointed out that angular measure is in terms of radians (π radians = 180 degrees). Thus, in example 4 above, THETA will be a number in the range zero to π. In example 5, it is apparent that variable ANGLE is expressed in degrees, but is converted to radians by dividing by the appropriate constant. Owing to the definition of the cosine function, variable COSINE will contain a real result in the range -1.0 to 1.0, after execution of the statement in example 5. Example 5 also shows that an argument may be an expression.

Library Function	Definition	Type of Argument	Type of Result
ABS(a)	$\lvert a \rvert$	Real	Real
AIMAG(a)	Obtain imaginary part of complex argument	Complex	Real
AINT(a)	Truncation	Real	Real
ALOG(a)	$\log_e(a)$	Real	Real
ALOG10(a)	$\log_{10}(a)$	Real	Real

FIGURE 5.5 Library functions

FIGURE 5.5—*continued*

Library Function	Definition	Type of Argument	Type of Result
AMAX0($a_1, a_2, \ldots$)	max($a_1, a_2, \ldots$)	Integer	Real
AMAX1($a_1, a_2, \ldots$)	max($a_1, a_2, \ldots$)	Real	Real
AMIN0($a_1, a_2, \ldots$)	min($a_1, a_2, \ldots$)	Integer	Real
AMIN1($a_1, a_2, \ldots$)	min($a_1, a_2, \ldots$)	Real	Real
AMOD(a_1, a_2)	a_1(mod a_2)	Real	Real
ATAN(a)	arctan(a)	Real	Real
ATAN2(a_1, a_2)	arctan(a_1/a_2)	Real	Real
CABS(a)	Absolute value of complex number	Complex	Complex
CCOS(a)	cos(a)	Complex	Complex
CEXP(a)	e^a	Complex	Complex
CLOG(a)	$\log_e(a)$	Complex	Complex
CMPLX(a_1, a_2)	$a_1 + a_2 \sqrt{-1}$	Real	Complex
CONJG(a)	Obtain conjugate of complex argument	Complex	Complex
COS(a)	cos(a)	Real	Real
CSIN(a)	sin(a)	Complex	Complex
CSQRT(a)	$\sqrt{a}$	Complex	Complex
DABS(a)	$\lvert a \rvert$	Double	Double
DATAN(a)	arctan(a)	Double	Double
DATAN2(a_1, a_2)	arctan(a_1/a_2)	Double	Double
DBLE(a)	Express single precision argument in double precision form	Real	Double
DCOS(a)	cos(a)	Double	Double
DEXP(a)	e^a	Double	Double
DIM(a_1, a_2)	a_1−min(a_1, a_2)	Real	Real
DLOG(a)	$\log_e(a)$	Double	Double
DLOG10(a)	$\log_{10}(a)$	Double	Double
DMAX1($a_1, a_2, \ldots$)	max($a_1, a_2, \ldots$)	Double	Double
DMIN1($a_1, a_2, \ldots$)	min($a_1, a_2, \ldots$)	Double	Double

FIGURE 5.5—*continued*

Library Function	Definition	Type of Argument	Type of Result
DMOD(a_1, a_2)	a_1 (mod a_2) *	Double	Double
DSIGN(a_1, a_2)	Sign of a_2 times $\mid a_1 \mid$	Double	Double
DSIN(a)	$\sin(a)$	Double	Double
DSQRT(a)	$\sqrt{a}$	Double	Double
EXP(a)	e^a	Real	Real
FLOAT(a)	Conversion from integer to real	Integer	Real
IABS(a)	$\mid a \mid$	Integer	Integer
IDIM(a_1, a_2)	$a_1 - \min(a_1, a_2)$	Integer	Integer
IDINT(a)	Sign of a times largest integer $\leqslant \mid a \mid$	Double	Integer
IFIX(a)	Conversion from real to integer	Real	Integer
INT(a)	Sign of a times largest integer $\leqslant \mid a \mid$	Real	Integer
ISIGN(a_1, a_2)	Sign of a_2 times $\mid a_1 \mid$	Integer	Integer
MAX0$(a_1, a_2, \ldots)$	$\max(a_1, a_2, \ldots)$	Integer	Integer
MAX1$(a_1, a_2, \ldots)$	$\max(a_1, a_2, \ldots)$	Real	Integer
MIN0$(a_1, a_2, \ldots)$	$\min(a_1, a_2, \ldots)$	Integer	Integer
MIN1$(a_1, a_2, \ldots)$	$\min(a_1, a_2, \ldots)$	Real	Integer
MOD(a_1, a_2)	a_1 (mod a_2) *	Integer	Integer
REAL(a)	Obtain real part of complex argument	Complex	Real
SIGN(a_1, a_2)	Sign of a_2 times a_1	Real	Real
SIN(a)	$\sin(a)$	Real	Real
SNGL(a)	Obtain most significant part of double-precision argument	Double	Real
SQRT(a)	$\sqrt{a}$	Real	Real
TANH(a)	$\tanh(a)$	Real	Real
IARFLT(i)	Determines if an arithmetic fault has occurred. The indicator is turned off when tested. Returns -1 if there is a fault; returns 0 if there is no fault.	Dummy **	Integer

FIGURE 5.5—*continued*

Library Function	Definition	Type of Argument	Type of Result
IDVCHK(*i*)	Determines if a divide fault has occurred. The indicator is turned off when tested. Returns -1 if there is a fault; returns 0 if there is no fault.	Dummy**	Integer
IEXFLT(*i*)	Determines if an exponent overflow has occurred. The indicator is turned off when tested. Returns -1 if there is a fault; returns 0 if there is no fault.	Dummy**	Integer
IFEOF(*i*)	Tests for end-of-file on unit *i* (section 9.5)	Integer	Integer
ISLITET(*i*)	Tests sense light *i* (*i*=1,24); turns it off if it was on. Returns -1 if error was detected, 0 if it was on, 1 if it was off.	Integer	Integer

*The function MOD(a_1,a_2) is defined as $a_1-[a_1/a_2]a_2$, where $[x]$ is the integer whose magnitude does not exceed the magnitude of x and whose sign is the same as x. Similarly for AMOD or DMOD.

**The *i* in the function reference is a dummy argument that is required but not used.

5.3 Statement Functions

The official nomenclature of this feature of Fortran is rather awkward. Let us therefore adopt the terminology AFDS, meaning Arithmetic Function Definition Statement.

An AFDS is a statement similar in appearance to an arithmetic assignment statement, but is in reality non-executable, and serves only to define a computational procedure in terms of symbolic arguments.

The general AFDS form is

$$f(a_1, a_2 \ldots a_n) = \text{arithmetic expression}$$

where

f is the function name and
a_i are the arguments.

The arguments are "dummy" variable names serving only to define the position and type of the required arguments. AFDS statements must precede the first executable statement and must follow specification and DATA (initialization) statements, if any. (Specification statements are DIMENSION and type declarations or COMMON or EQUIVALENCE statements; COMMON statements are discussed in a subsequent chapter.)

The use of AFDS is depicted in Figure 5.7, which is a program to produce geometric data for a square-based pyramid. Figure 5.7 is a solution of the problem illustrated in Figure 5.6. Note the agreement between output and the results of hand calculations; the user of a computer must always prove the validity of a program before relying on the program for large amounts of useable output.

Let $s = 5$, $h = 7$

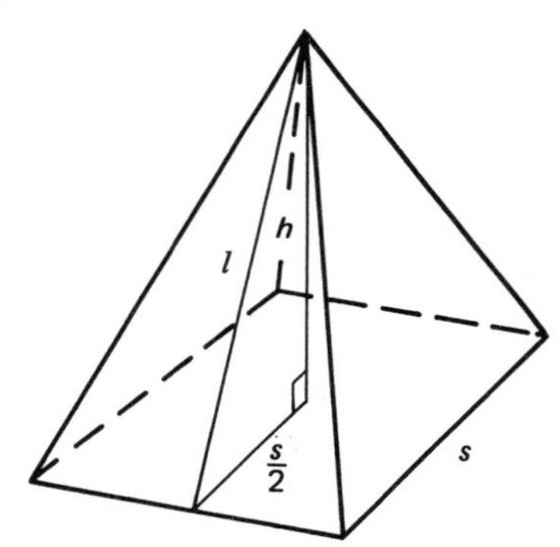

Where

h is perpendicular height
s is length of a side
l is "slant height"

Base Area $= s^2 = 25$

Volume $= \frac{1}{3} \cdot s^2 \cdot h = \left(\frac{25}{3}\right) \cdot 7 = \frac{175}{3} = 58.33$

Slant Height l $= \sqrt{\left(\frac{s}{2}\right)^2 + h^2} = \sqrt{\left(\frac{5}{2}\right)^2 + 7^2} = \sqrt{55.25} = 7.43$

Area of surface (4 sides) $= 4\left(\frac{1}{2} \cdot l \cdot s\right) = 2\,s\,l \doteq 74.3$

FIGURE 5.6 Pyramid geometry hand calculation

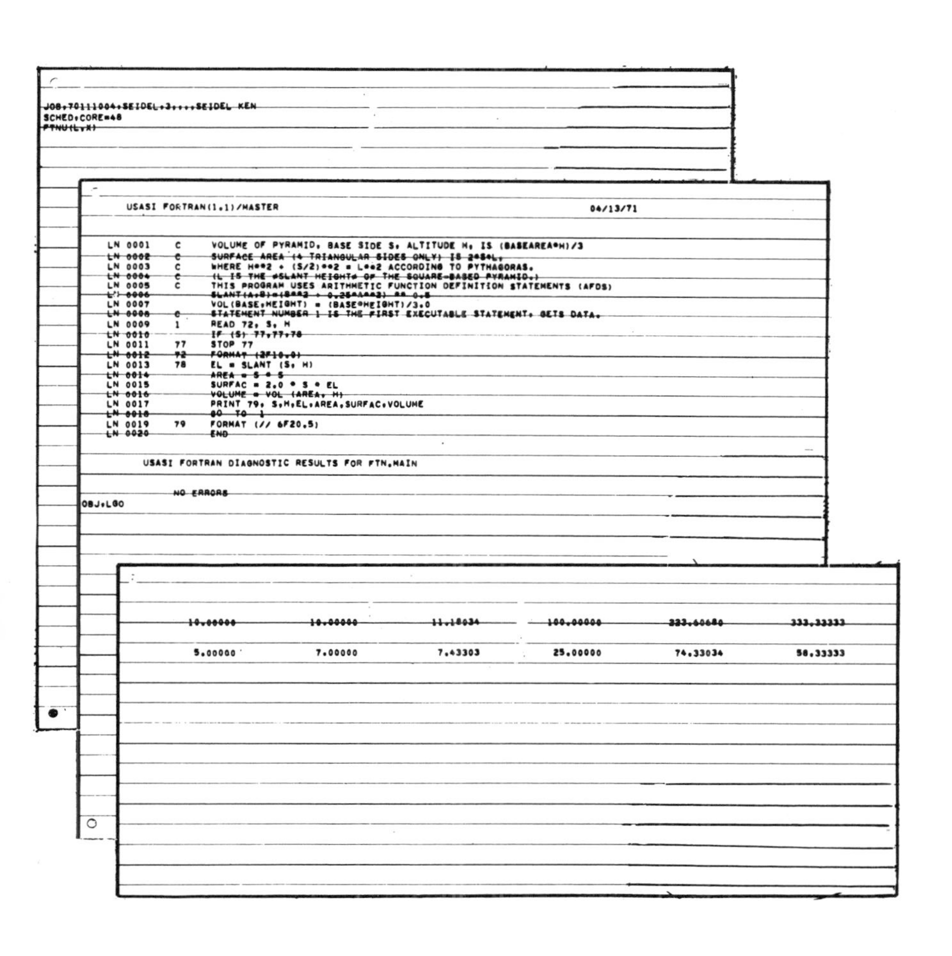

```
JOB,70111004,SEIDEL,3,,,,SEIDEL KEN
SCHED,CORE=48
FTNU(L,X)

     USASI FORTRAN(1.1)/MASTER                                    04/13/71

   LN 0001   C     VOLUME OF PYRAMID, BASE SIDE S, ALTITUDE H, IS (BASEAREA*H)/3
   LN 0002   C     SURFACE AREA (4 TRIANGULAR SIDES ONLY) IS 2*S*L,
   LN 0003   C     WHERE H**2 + (S/2)**2 = L**2 ACCORDING TO PYTHAGORAS.
   LN 0004   C     (L IS THE #SLANT HEIGHT# OF THE SQUARE-BASED PYRAMID.)
   LN 0005   C     THIS PROGRAM USES ARITHMETIC FUNCTION DEFINITION STATEMENTS (AFDS)
   LN 0006         SLANT(A,B)=(B**2 + 0.25*A**2) ** 0.5
   LN 0007         VOL(BASE,HEIGHT) = (BASE*HEIGHT)/3.0
   LN 0008   C     STATEMENT NUMBER 1 IS THE FIRST EXECUTABLE STATEMENT, GETS DATA.
   LN 0009   1     READ 72, S, H
   LN 0010         IF (S) 77,77,78
   LN 0011   77    STOP 77
   LN 0012   72    FORMAT (2F10.0)
   LN 0013   78    EL = SLANT (S, H)
   LN 0014         AREA = S * S
   LN 0015         SURFAC = 2.0 * S * EL
   LN 0016         VOLUME = VOL (AREA, H)
   LN 0017         PRINT 79, S,H,EL,AREA,SURFAC,VOLUME
   LN 0018         GO TO 1
   LN 0019   79    FORMAT (// 6F20.5)
   LN 0020         END

        USASI FORTRAN DIAGNOSTIC RESULTS FOR FTN.MAIN

           NO ERRORS
OBJ,LGO

      10.00000     10.00000     11.18034     100.00000     223.60680     333.33333

       5.00000      7.00000      7.43303      25.00000      74.33034      58.33333
```

FIGURE 5.7 Pyramid program run

In Figure 5.7, AFDS are used to define the SLANT and VOL functions. The SLANT function has two arguments arbitrarily designated by dummy names A and B, corresponding to *s* and *h* in Figure 5.6. Note that use of the SLANT function in line 0013 of the program in Figure 5.7 supplies the values of variables

S for A

and

H for B.

The VOL function has two dummy arguments, BASE and HEIGHT. The use of this function, in line 0016 of Figure 5.7, supplies the values of variables

AREA for BASE

and

H for HEIGHT.

The program reads one data card, calculates and outputs the appropriate data, and recycles for another data card, stopping only when zero or negative data for *s* is encountered. In effect, each expression or variable in the parenthesized argument list of an AFDS reference is substituted into the prototype AFDS statement instead of the corresponding "dummy" argument variable. Use of a particular name as a dummy argument does not preclude use of the same name as a variable, or as a dummy argument in another AFDS.

Apart from dummy arguments, the expression in an AFDS may contain only:

Constants

Variable references

Library function references

AFDS references, provided the function is defined previously

External function references[10]

10. Refer to the chapter on subprograms.

Exercises

1. Write a program that can reverse the data in a one-dimensional array. Assuming that the array has 72 integer members, the program should read them in from cards (12 numbers per card, in columns 1-72), print the array, rearrange it and then print out the array. The rearrangement should be such that the original value I_1 ultimately resides in I_{100}; the original I_2 ultimately resides in I_{99}, and so forth.

2. Write a program to read in a 2-dimensional matrix of real numbers. The matrix should have 4 rows and 4 columns. After reading the matrix, print it out as 4 lines, double spaced, with each line representing a row.

   Now form the transpose of the original matrix and print out its rows. For ease in testing, each input value should be unique.

3. Four men are in joint possession of a pile of coconuts. While the other three are asleep, one awakes and divides the pile of coconuts into four equal piles, with one left over which he gives to a passing monkey. Thereupon, he hides one of the four piles, and then combines the three remaining piles and returns to his bunk.

   In separate turns, each of the other three men awakes and performs the same task; the monkey receiving the "odd remaining" coconut in each case.

   In the morning, when all four are awake, the group divides the remaining single pile (considerably diminished as a result of the nightly skulking) into four equal piles, with a single leftover coconut for the ever-present monkey.

   What is the minimum number of coconuts in the original pile?

6

DECK SET-UP REQUIREMENTS

6.1 MASTER Operating System

For every general purpose digital computer, a software "operating system" is provided. One of its aims is to schedule jobs and supervise the initiation of other processors, such as Fortran and COBOL compilations, as specified by control cards that precede each job.

In order to run a Fortran job in the CDC 3170/3300/3500 computers throughout the California State College system using the MASTER Operating System, we organize the job deck as presented in this section. The deck order is:

1. $JOB card
2. $SCHED card
3. $FTNU card
4. Fortran source statements
5. FINIS card
6. $OBJ card
7. Input data, if any
8. End-of-job control card.

Compile and execute

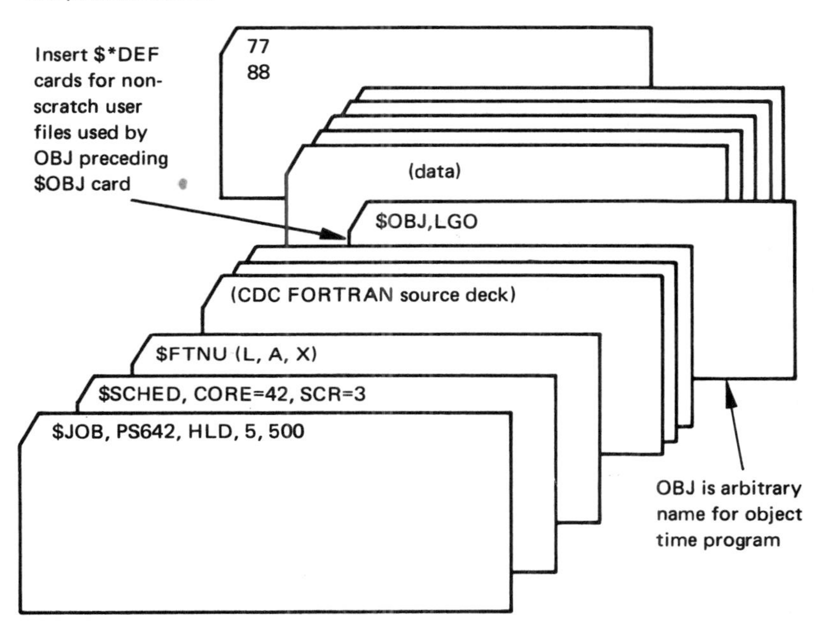

Compile and punch

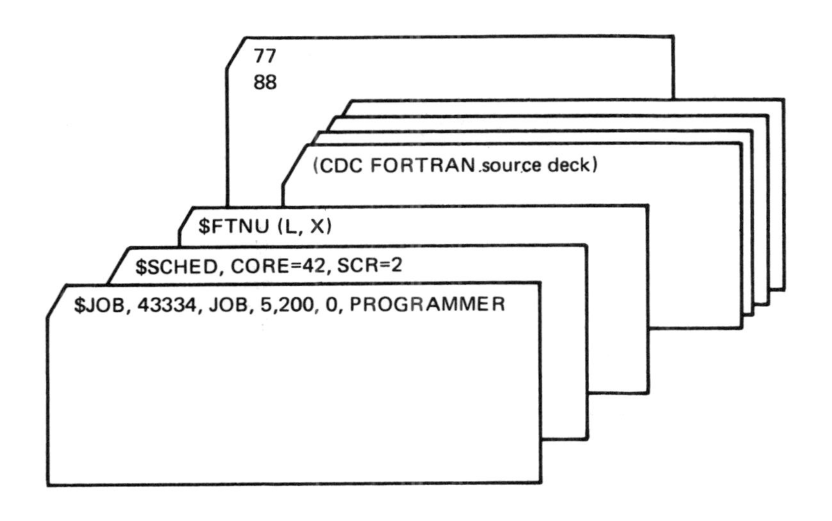

FIGURE 6.1 Typical MASTER Fortran deck set-ups

6.2 $JOB card

The $JOB card has the general form

%JOB,*j*,*n*,*t*,*q*,*p*,*s*,name

and is punched beginning in column 1. Except for name, there are no blanks in the above information. Parameters *j*, *n*, *t*, *q*, *p*, and *s* are as defined below; name is in the form lastname firstname.

j is an 8 digit job number assigned to the user.

n is an 8 character (or less) job name; it may be the first 8 characters of the user's last name, or the entire last name (if not over 8 characters).

t is the maximum time limit in minutes.

q is the maximum number of printed lines of output.

p is the maximum number of punched cards to be produced in the run.

s is normally not present, which is indicated by a comma immediately following the comma that follows *p*; *s* is otherwise the job name of another job that must be run before this one.

Only *j*, *n*, and name are required. If the other parameters are null, indicated by successive commas, standard values are assumed, which is normally sufficient for student test programs.

6.3 $SCHED Card

This control card specifies the machine resources required during compilation and/or execution. In the case of student test programs, where no external storage devices other than card reader, card punch, or printer are required, the following is a sufficient $SCHED card:

$SCHED,CORE=48,SCR=3

6.4 $FTNU Card

This card calls upon the USASI Fortran compiler, and precedes the source statements. The typical form is:

$FTNU (L,X)

where the parenthesized parameters have the following interpretations:

L: Source program listing is desired.

X: Execution is desired after compilation.

S: Integer and Logical variables are compiled as one-word entities (making maximum integer = $2^{23}-1$).

A: Object code generated by the compiler will be listed with the source program.

R: A cross-reference listing is desired.

D: The user requests diagnostic message notification for any non-USASI source language usage.

O: The user requests that the compiler attempt to optimize the generated object program for greater execution time efficiency.

6.5 FINIS and $OBJ Cards

Following the last source statement (which must be END), a card having the word

FINIS

punched in columns 10–14 must appear.

The next control card should be

$OBJ,LGO

which initiates execution of the user's object program from the standard load-and-go file created during compilation.

6.6 End-of-job Card

The aforementioned $OBJ card must be followed by the necessary data cards required as input to the program. The amount and form of these data cards (if any) is a matter controlled by the designer of the program. The last card in every job deck, however, must contain only two "7-8" punches in columns 1 and 2.

6.7 Diagnostic Messages

If a program contains errors, such as the violation of Fortran language rules, the compiler prints "diagnostic" messages and prevents the execution of the faulty object program from taking place. Figure 6.2 illustrates the printout results from the computer in such a case.

```
JOB,70111004,SEIDEL,3,,,,SEIDEL KEN
SCHED,CORE=48
FTNU(L,X)


        USASI FORTRAN(1,1)/MASTER


     LN 0001      1        READ 10,X
     LN 0002               IF (X) 2,9,2
     LN 0003      10       FORMAT (E10.4)
     LN 0004      2        PRINT 11,X
     LN 0005      11       FORMAT (1H0, E20.8)
     LN 0006               GO TO 1
     LN 0007      9        STOP 9
     LN 0008               END


        USASI FORTRAN(1,1)/MASTER

           USASI FORTRAN DIAGNOSTIC RESULTS FOR FTN.MAIN

        LINE    S    ERRNUM                        MESSAGE

        0007    F     9012       ILLEGAL DIGIT IN OCTAL CONSTANT.
OBJ,LGO
EXECUTION DELETED
```

FIGURE 6.2 Compiler output from faulty program

6.8 MSOS

MSOS is a mass-storage-based operating system implemented for CDC 3100, 3150, 3200, 3300, and 3500 computer systems. MSOS pre-dates the MASTER operating system and is generally considered to be the less sophisticated of the two. For example, a non-USASI Fortran compiler under MSOS is less comprehensive, lacking certain "Fortran IV" features such as function definition statements, named common, complex variables, and so forth. In essence, the Fortran available under MSOS is an extended version of Fortran II.

Operator's console, CDC 3300 computer system

The sample program shown in Figure 2.3 (real roots of a quadratic equation) was run under MSOS, using the control cards appropriate to that operating system. The printed results of compilation and execution are shown in Figure 6.3. The elapsed time of the run was 7 seconds.

Control Data 3300

```
                SEQUENCE,003,  TIME    1349           DATE    022371
                JOB,10001000,PYR  SEIDEL KEN ,,,ND
                FORTRAN,L,X


                                        31/32/3300 FORTRAN (3.1)/MSOS   02/23/71
          PROGRAM REALROOT
      C   CALCULATE ROOTS OF A QUADRATIC EQUATION.
      C   IF  ROOTS ARE NOT REAL (DISCRIMINANT NEGATIVE),
      C   STOP IMMEDIATELY.
      C   1. SET UP COEFFICIENT VALUES, THEN COMPUTE  DISCRIMINANT.
          A = 1.5
          B = 9.8
          C = 0.63
          DISCRM = B ** 2 - 4.0 * A*C
      C   2. TEST FOR NEGATIVE DISCRIMINANT.
          IF  (DISCRM) 16,14,14
     14   SURD = DISCRM ** 0.5
          D = 2.0*A
          X1 = (-B+SURD)/D
          X2 = -(B+SURD)/D
          PRINT  15, A,B,C,X1,X2
     15   FORMAT(F9.3,F9.3,F9.3,E18.9,E18.9)
     16   STOP   16
          END

   COMPILED LENGTHS OF REALROOT - P = 00143   C = 00000   D = 00000
LOAD,56
RUN,,NM
A ELAPSED TIME 00/00/11
   1.500    9.800    .630  -6.493102619E-02  -6.468402307E 00
 A FTN 061 ( STOP 00016)
```

FIGURE 6.3 Program run under MSOS

The following deck set-up is required for MSOS. In the explanations, the symbol ∇ signifies punches in both rows 7 and 9 of card column 1.

1. Sequence card:

   ∇SEQUENCE,*n*

   where *n* is any 3-digit number.

2. Job card:

   ∇JOB,user #,job name, , ,ND

   where job name is any alphabetic designator including embedded blanks, if desired.

3. Compiler call card:

   ∇FORTRAN,L,X

   where L and X options request listing and load-and-go, respectively.

4. Program name card:

   PROGRAM name

   where name is any 8-character "variable" name given to the entire program. This is a Fortran style statement, so the word PROGRAM begins in column 7.

5. Remainder of first Fortran program: if there are several subprograms, the FUNCTION or SUBROUTINE statement of each subprogram must be preceded by a blank card. Each program unit has its own END card. A single FINIS card (col. 10) follows the last END card.

6. ∇LOAD,56

   where 56 is the logical unit number of the standard "load-and-go file" (which is built up by the compiler).

7. ∇RUN,NM

   where the NM option requests that no "map" be produced.

8. Any card input data required by the program.

9. End-of-job card (see section 6.6).

Exercises

1. Write a polynomial curve plotting program. The polynomial

$$y = f(x) = c_{n+1} \cdot x^n + c_n \cdot x^{n-1} + \ldots + c_2 \cdot x + c_1$$

can be determined by input of the integer n followed by input of the n+1 real coefficients. Allow for $1 \leqslant n \leqslant 9$. (Your program should check that the n input conforms to this restriction. If not, print an appropriate diagnostic message and continue with the next set of data.) Test for the special "flag" value $n = 999$, to be used as the program termination trigger.

After receipt of the polynomial's degree and coefficients, read the real values XSTART and XSTEP, plus the integer M.

XSTART is the leftmost point on the X-axis.
XSTOP is the rightmost point on the X-axis.
M is the number of points (along the X-axis) at which the polynomial $y = f(x)$ is to be evaluated. $M \leqslant 101$.

Now, the program computes the M Y-values of the curve for the X values

XSTART
XSTART + XSTEP
XSTART + 2 * XSTEP, etc.

Output of the program shall be a plot of the M points of the curve, along with suitably labeled X and Y axes.

The X axis should be across the output page in the horizontal direction. The Y axis should be at the left of the page, and must take at least 41 lines and at most 101 lines in the vertical direction. The axes should be printed as lines of periods; the points of the curve should be printed as circles (letter O). Grid lines should be printed at every tenth point, in either direction.

Whenever a point of the plotted curve coincides with a point on the grid, the circle character should prevail.

7

TIME-SHARING FORTRAN

7.1 KRONOS III Time-Sharing System

Today's computers are communications-oriented. The complex and comprehensive operating systems have the capability of managing many jobs on a concurrent (multi-programming) basis.

Instead of the old-fashioned way of doing things, where one job at a time occupied the entire computer, regardless of how inefficiently, today's computers are programmed to respond to many concurrent users, some of who access the computer from a terminal (such as a Teletype keyboard device).

Because of its large storage capacity and high-speed data handling capability, the computer can be equipped (through an Operating System) to respond to many terminal-based users, and each one is given sufficiently rapid responses to satisfy his requirements.

One of the great advantages of time-sharing is freedom from the tyranny of the punched card deck, and the attendant long turn-around times usually associated with computer jobs run in a batch processing environment.

With access to the computer provided through time-sharing terminals, the experimental problem solver gains very rapid feedback from the computer, and can therefore make more productive use of his time, since results are obtained nearly instantaneously.

In this chapter, we will present enough information about the KRONOS III time-sharing to enable the Fortran programmer to create, maintain, and execute Fortran programs from a Teletype terminal.

First, however, we point out that the Fortran system available under KRONOS III is not as comprehensive as the USASI Fortran available under the MASTER batch processing system. The principal time-sharing Fortran limitations (which are fairly typical of time-sharing system restrictions) are the exclusion of double precision, complex, and logical data types, and limiting integer variables to one 24 bit word.[11]

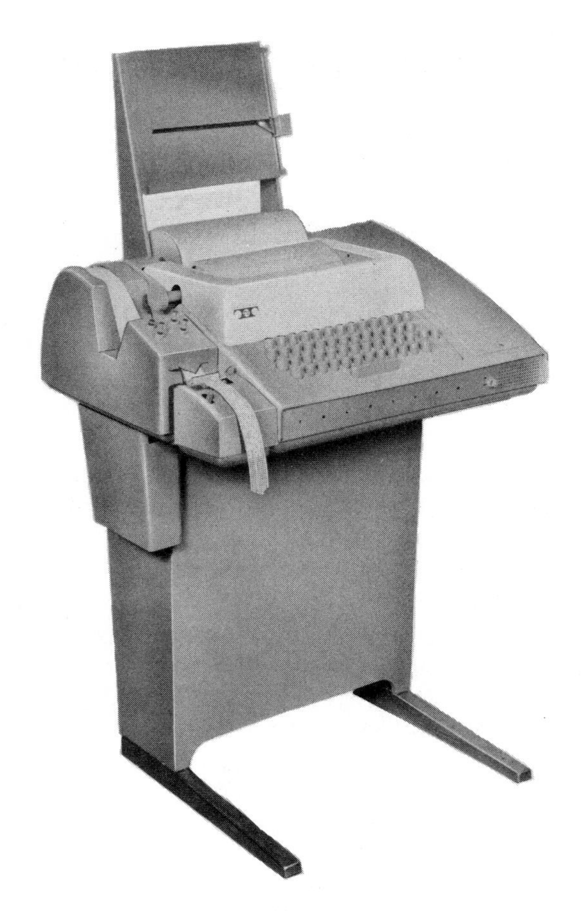

Teletype Model 33 ASR

11. A variant version of Fortran called FTN is available under the KRONOS III time-sharing system. Since FTN contains numerous differences from FORTRAN, we are confining our concern here to the more standard compiler, FORTRAN, under KRONOS III.

Another difference is that printing action does not interpret the first output character of each line as the print control character. Instead, all output characters are printed. Line width is restricted to 72 characters on the teletype.

Teletype Model 33 ASR keyboard

7.2 Log-in Procedure

The user "turns on" the terminal, or dials the computer access telephone number; the computer responds by printing

PLEASE IDENTIFY YOURSELF:

The user responds by typing his ID code in the format

aaaapppppp;nnnnnn (CR)

where

the *a*'s represent a 4-digit account number assigned to each user,
the *p*'s represent a password of up to 6 characters,
the *n*'s represent a user name of up to 6 characters, and
(CR) represents the carriage return (RET key)

If any of the above elements is incorrect (or not previously certified at the computer center as permissible codes), the identification request is repeated, and the user must supply the correct response within 90 seconds.

When the ID is accepted, the system types the data, time, phone line numbers, and then

WHAT SYSTEM?

User responses can be:

EXEC (CR)
EDIT (CR)
FORTRAN (CR)
LØGØUT (CR)
plus others not of concern to the Fortran user.

The LOGOUT response is used to sign off the user and disconnect the terminal.

Throughout his work at the teletype terminal, the user may correct key-in errors by using any of three special control characters. These control characters are produced by holding the CTRL key down and then pressing either the C, W, or L letter key.

CTRL-C This pair of keys, which prints the character ← , deletes the previous single character.

CTRL-W This pair of keys, which prints as ↑ , deletes all previous characters to a leading blank. It is most often needed to delete a preceding word.

CTRL-L This pair of keys, which prints as \ , deletes the entire line of text (provided the CR terminating the line has not been given yet).

The notation for these three control characters is sometimes given as C^c, W^c, or L^c, respectively.

At any time, the user may terminate the current operation by using the "Escape" key (labeled Alt Mode on the keyboard). This causes whatever system is currently in operation to print its prompt character in readiness to accept a command. The prompt characters are

- \- for EXEC
- * for EDIT
- \+ for FORTRAN

This Escape operation is useful whenever the user seems to have lost control as when, for example, a program execution runaway (loop) occurs. It is also useful when it is desired to terminate execution of a program that has no built-in method of stopping. An example of this is shown in Figure 7.7, panel B. Note: the Escape (Alt Mode) key does not print anything.

7.3 EXEC

Responding EXEC makes it possible to type a variety of commands, such as the following, after being "prompted" by EXEC's printing of the hyphen symbol. Each command below is followed immediately by CR.

FILES	Prints the name of each file saved for the user.
NEWS	Prints current announcements of general interest to all users.
RENAME filename AS filename	Alters the name of a file.
LOGOUT	Logs the user off the system and disconnects the terminal.
CONTINUE	Returns the user to the system in use immediately before EXEC.
DELETE filename	Eliminates a previously saved file.
EXIT	Exits from EXEC and returns to the request WHAT SYSTEM?
MEMORY	Prints a memory map of all pages in core memory and the total number of computer words currently assigned to the user.
CLEAR	Releases all pages of core memory–used to change object programs in Fortran.
STORAGE	Prints information about the user's disk storage including maximum storage available, the amount in use, and the charge units accrued since the last billing period.[12]
TIMES	Prints the date and time the user logged in and both the elapsed (connect) time (in hours and minutes) and the computer time used (in hours, minutes, and seconds). These figures are given for both the current session and the current billing period.
DATE	Prints the current date and time of day.

12. The billing period and charge units are not of concern to the student user.

7.4 EDIT

EDIT is a general purpose text editor, designed for the creation and alteration of symbolic text, including source programs. The asterisk (*) is the "prompt" character from EDIT, which indicates that EDIT is waiting for commands from the user.

To create a Fortran program, the first command should be to set the "tab" for position 7:

*TAB 7(CR)

If the user forgets to set the tab, he must strike the space key in order to position the input mechanism correctly. For example, a statement having no statement number would have to begin with 6 spaces in order to place the first non-blank character in the position corresponding to "column" 7. When a statement number is given, there must be enough following spaces to reach "column" 7 before typing the statement.

The next command should be

*APPEND (CR)

to indicate that a program creation sequence is about to follow.

At this point, the desired Fortran statements are entered as a series of discrete lines, each followed by (CR) .

When typing a particular line, the following rules are applicable (provided TAB was set to 7):

1. If the line has no statement number, strike the tab key and then type in the statement.
2. If there is a statement number, type it, then tab, then the statement.
3. For continuation lines, strike the space key 5 times; then the appropriate digit for continuation, followed by the source statement's continuation.

Upon completion of the program (don't forget the END statement), the R key is depressed at the same time as the control (CTRL) key is held down, in order to terminate the APPEND operation. At this point, EDIT is still operating, and the program just keyed-in is in the "Main Buffer."

The next operation should be to save the main buffer as a file:

*SAVE filename (CR)

where filename is up to 9 non-blank characters (letters and digits only are recommended). Each user can have up to 39 files.

The SAVE operation elicits the query NEW FILE? to which the user should respond YES (CR) .

In a later section, we will discuss the EDIT functions that permit the user to modify a program. But now, we are ready to compile the program just saved, so we issue our last response to EDIT:

*EXIT (CR) .

7.5 KRONOS III Fortran

After our exit from EDIT, KRONOS III asks WHAT SYSTEM? to which we respond

FORTRAN (CR) .

This causes a question from the system: NEW OR OLD–to which we respond

NEW (CR) .

Next, the system prints SOURCE INPUT FROM: to which we respond

filename (CR)

using any filename saved previously.

The next printout from the system is OLD FILE? to which we respond

YES (CR) .

The Fortran system now prints BINARY OUTPUT TO: and our response should be

filename (CR)

in which we designate a filename different from any of those previously saved. (The FILES printout obtained from an earlier EXEC session is useful in aiding the user in his choice of a unique filename.) The binary output involved in this step is the object program.

The final query from the system is LISTING ON: and our response should be

TELETYPE (CR) .

The Fortran compilation now takes place, and the program execution ensues.

Within the Fortran language itself, there are several variants added for the convenience of the terminal-based user.

To accept data from the keyboard, the formatted READ statement form

READ format, list

can be used; the user must take great care to input exactly as many characters per field as described in the format, which can be inconvenient in many cases. Therefore an unformatted read statement is available, where the successive data values input at execution time will be simply separated by commas. (See Figure 7.7.) The form of this statement is

READ, list

and the comma is required.

Similarly, there is an unformatted output facility:

PRINT, list

that is convenient for use in program situations where exactly tailored print layouts are not required.

Immediately upon the completion of the compilation stage, an object program enters the execution stage, automatically, with no explicit message or signal noting this fact. Therefore, it is strongly recommended that the user construct his programs so that they "cue" him prior to the need for specific input. Example:

```
    PRINT 101
101 FORMAT (19HNOW KEY-IN X AND N.)
    READ, X, N
```

When the program containing the above fragment is executing, the message

NOW KEY-IN X AND N.

will indicate what the user must do to continue the program action. Without such a "cue," the situation can become ambiguous.

At the end of execution, the Fortran prompt character (+) is typed. The next response should be EXIT (CR). If the program does not compile correctly due to a source statement error, the program listing will contain an indication of the error, execution is bypassed, and the prompt character is typed. In this case, the program must be modified, and the user should exit.

Users of KRONOS III Fortran may specify character literals in a format statement or output list. This facility has the advantage of being simpler than the *n*H format specification.

Example:

```
      PRINT, "STATISTICAL FORECAST"
      PRINT 76, N, X
76    FØRMAT (/ ,"CASE NO.",I4,E17.9)
```

When a READ statement is executed, the teletype bell is sounded, and the system waits for input to be keyed in. During the key-in process, a keying error can be corrected by using C^c (the CTRL-C combination of keys) in order to delete the immediately preceding character of a line.

7.6 Program Changes Using EDIT

If a saved source program (file) must be changed because an error is detected during compilation, the user should exit from Fortran and re-enter EDIT. At this time, one should immediately set the tab to 7, as during the program creation session.

Now, however, the user wants to load the program file and change parts of it. To do so, he uses commands to:

1. Load a previously saved file back into the "main" buffer;
2. Print all or part of the main buffer;
3. Change a line or contiguous set of lines;
4. Insert new lines before a designated line;
5. Delete a line or contiguous set of lines.

In response to the *prompting from EDIT, the operation

*LOAD filename (CR)

brings an old file into the main buffer, in order to prepare for modifications. (Recall that we start a program creation sequence with *APPEND.)

To delete a line, determine its number (a) and respond

To delete lines a through b, respond

*a,bD (CR)

At any point in time, including just after deleting some lines, the program can be printed by responding

*PRINT (CR)

Line a can be printed by responding

*a/ (CR)

while lines a through b are printed by

*a,b/ (CR)

To change a specific line, use

*aC (CR)

followed by new text lines, and terminated by the simultaneous use of the CTRL and R keys, as during the program creation (*APPEND) session.

To insert new lines, determine the line number (a) just before the point at which the inserts should go, and use the response

*aINSERT (CR)

followed by the desired lines and terminated by the simultaneous use of the CTRL and R keys.

After all corrections have been made, it is good practice to print the entire buffer and inspect it prior to the SAVE operation. After saving it (usually by an old filename), the user will exit from EDIT, and enter Fortran in order to compile and execute.

Figures 7.1 through 7.5 illustrate the development of a Fibonacci series generator program. The first compilation revealed an error (statement number Y); this was corrected, recompiled, and executed. In Figure 7.5, note the free-format output that execution of FIBØBJ produces.

```
WHAT SYSTEM? EDIT

KRONOS III EDIT
*TAB 7
*APPEND
Y      I=Ø
       J=1
       LIM=5ØØØØ
       K= I+J
       AK=K
       AJ = J
       R=AK/AJ
       PRINT, K,R
       I=J
       J=K
       IF(K-LIM)1,1,7
7      STOP 747
       END


*4C
1      K=I+J

*SAVE FIB
NEW FILE? YES
*EXIT

WHAT SYSTEM? FORTRAN

NEW OR OLD--NEW
SOURCE INPUT FROM: FIB
OLD FILE? YES
BINARY OUTPUT TO: FIBOBJ
NEW FILE? YES
LISTING ON: TELETYPE
```

FIGURE 7.1 Developing program FIB

```
Y       I=0
SYNTAX
        J=1
        LIM=50000
1       K=I+J
        AK=K
        AJ = J
        R=AK/AJ
        PRINT, K,R
        I=J
        J=K
        IF(K-LIM)1,1,7
7       STOP 747
        END

SUBPROGRAMS

  R.CLSALL

PROGRAM ALLOCATION

  00051   J     00052   LIM     00053   K     00054   I
  00055   AK    00057   AJ      00061   R

00000000 00001    ERRORS


TOTAL ERRORS 00001


+EXIT
```

FIGURE 7.2 First compilation of FIB

```
PLEASE IDENTIFY YOURSELF: 1148111004;SEIDEL

KRONOS III CALLIN TIME IS:

11:35 A.M.  DEC 7, 1970

PHONE LINE 1027
TIME SHARING WILL GO DOWN AT 1:30 TODAY. IF YOU HAVE ANY
PROBLEMS BRING YOUR OUTPUT TO ENG 121 OR CALL THE TIME-
SHARING OPERATOR AT 1484.

WHAT SYSTEM? EDIT

KRONOS III EDIT
*LOAD FIB
OLD FILE? YES

*TAB 7
*PRINT
Y      I=0
       J=1
       LIM=50000
1      K=I+J
       AK=K
       AJ = J
       R=AK/AJ
       PRINT, K,R
       I=J
       J=K
       IF(K-LIM)1,1,7
7      STOP 747
       END
*1C
       I = 0

*SAVE FIB
OLD FILE? YES
*EXIT
```

FIGURE 7.3 Correcting FIB

```
WHAT SYSTEM? FORTRAN

NEW OR OLD--NEW
SOURCE INPUT FROM: FIB
OLD FILE? YES
BINARY OUTPUT TO: FIBOBJ
OLD FILE? YES
LISTING ON: TELETYPE


      I = Ø
      J=1
      LIM=5ØØØØ
1     K=I+J
      AK=K
      AJ = J
      R=AK/AJ
      PRINT, K,R
      I=J
      J=K
      IF(K-LIM)1,1,7
7     STOP 747
      END

SUBPROGRAMS

  R.CLSALL

PROGRAM ALLOCATION

  ØØØ54  I     ØØØ55  J     ØØØ56  LIM        ØØØ57  K
  ØØØ6Ø  AK    ØØØ62  AJ    ØØØ64  R
```

FIGURE 7.4 Recompiling FIB as FIBØBJ

```
 1, 1
 2, 2
 3, 1.5
 5, 1.66666666
 8, 1.6
 13, 1.625
 21, 1.61538461
 34, 1.61904761
 55, 1.61764705
 89, 1.61818181
 144, 1.61797752
 233, 1.61805555
 377, 1.61802575
 610, 1.61803713
 987, 1.61803278
 1597, 1.61803444
 2584, 1.61803381
 4181, 1.61803405
 6765, 1.61803396
 10946, 1.61803399
 17711, 1.61803398
 28657, 1.61803399
 46368, 1.61803398
 75025, 1.61803398
STOP      747

+EXIT

WHAT SYSTEM? EXEC

-LOGOUT

CONNECT: 0 HRS. 7 MINS.
COMPUTE: 0 HRS. 0 MINS. 2 SECS.
```

FIGURE 7.5 Executing FIBØBJ

7.7 Executing Old Object Programs

Figure 7.6 shows a KRONOS III terminal printout obtained by executing a previously saved Fortran object program. Note the output "cue" text (SUPPLY PRIME LIMIT), and the value 456 keyed in by the user. All subsequent printing emanates from the execution of object program OBJPRIM2, until the program STOP occurs.

```
PLEASE IDENTIFY YOURSELF: 1148111004;SEIDEL

KRONOS III CALLIN TIME IS:

8:31 P.M.  DEC 1, 1970

PHONE LINE 1005

WHAT SYSTEM? DEXEC

WHAT SYSTEM? EXEC

-FILES
5
PRIME
OBJPRIME
OBJPRIM2
QUADROOTS
QUADOBJ

-EXIT

WHAT SYSTEM? FORTRAN

NEW OR OLD--OLD
LOAD FROM BINARY FILE: OBJPRIM2
OLD FILE? YES

SUPPLY PRIME LIMIT
456
 PRIME NOS. IN RANGE 3-   456
     3
     5
     7
    11
    13
    17
    19
    23
    29
    31
    37
    41
    43
    47
     .
     .
   389
   397
   401
   409
   419
   421
   431
   433
   439
   443
   449
 STOP        0

+EXIT
```

FIGURE 7.6 Executing an old project program

Figure 7.7 shows the use of the CONTINUE and CLEAR commands during an extensive session at the computer terminal.

In Panel A, source file CHISQ is compiled and executed. Note that the program has no logical execution end–it recycles to statement number 1 after the result is printed. Also note the fact that the program "acknowledges" the 4 input values; this type of positive control on the part of the programmer is a practice well worth your imitating.

```
WHAT SYSTEM? FORTRAN

NEW OR OLD--NEW
SOURCE INPUT FROM: CHISQ
OLD FILE? YES
BINARY OUTPUT TO: CHIOBJ
NEW FILE? YES
LISTING ON: TELETYPE


1       READ,A,B,C,D
        PRINT 22,A,B,C,D
22      FORMAT(//4F10.4)
        EN=A+B+C+D
        X=ABS(A-C)
        Y=(X-1.0)**2
        UP=EN*Y
        DOWN=(A+C)*(B+D)
        CHI=UP/DOWN
        PRINT 33, CHI
33      FORMAT(/"CHI VALUE =", F11.5//)
        GO TO 1
        E N D


SUBPROGRAMS

  ABS          R.CLSALL

PROGRAM ALLOCATION

  00117  A          00121  B          00123  C          00125  D
  00127  EN         00131  X          00133  Y          00135  UP
  00137  DOWN       00141  CHI




5.,8.,8.,5.


    5.0000     8.0000     8.0000     5.0000

CHI VALUE =       .61538
```

FIGURE 7.7 An extensive session at the computer terminal (Panel A)

In Panel B, we see that the Fortran prompt character appeared as a result of the Escape (Alt Mode) key having been pressed by the terminal operator. Use of the RUN response re-initiated CHIOBJ execution. Later, another Escape was used to call EXEC. Note the use of CONTINUE command, and Fortran's re-sign-on message FORTRA. The next action keyed-in at the terminal is a request to LOAD ROOTS (note the use of 5 consecutive C^c character deletions, which print as ◄—'s). Upon loading ROOTS, there are now two main programs in memory. This ambiguity is resolved as shown in Panel C.

```
+RUN

4.4,5.5,6.6,3.3

    4.4000      5.5000      6.6000      3.3000

CHI VALUE =       .29455

12.,7.,11.,8.

   12.0000      7.0000     11.0000      8.0000

CHI VALUE =     0.00000

12.,7.,8.,11.

   12.0000      7.0000      8.0000     11.0000

CHI VALUE =       .95000

<ESCAPE>
+EXIT

WHAT SYSTEM? EXEC

-DELETE JOSE
OLD FILE? YES

-DELETE JOSEXX
OLD FILE? YES

-FILES
10
PRIME
FIB
PRIMOBJ
FIBOBJ
QUADROOTS
ROOTS
CHIOBJ
CHISQ
FMTEST
FMTOBJ

-CONTINUE
FORTRA

+LOAD QUADR←←←←←ROOTS
OLD FILE? YES

.MAIN.
MULTIPLY-DEFINED ENTRY POINT
+EXIT
```

FIGURE 7.7 (Panel B)

In Panel C, the user reverts to EXEC and issues the CLEAR command, eliminating the CHIOBJ program from user core memory. After one false step, he returns to Fortran, loads ROOTS, and obtains the solution to the desired quadratic equation. Finally, EDIT is called to provide another printed copy of the source file QUADROOTS, which has ROOTS as its object program.

```
WHAT SYSTEM? EXEC

-CLEAR

-CONTINUE

ILLEGAL COMMAND

-EXIT

WHAT SYSTEM? FORTRAN

NEW OR OLD--OLD
LOAD FROM BINARY FILE: ROOTS
OLD FILE? YES

2.,66.,4.
A=  2.00000   B= 66.00000   C=   4.00000

REAL ROOTS...   -6.071778E-02  -3.293928E+01


+EXIT

WHAT SYSTEM? EDIT

KRONOS III EDIT
*LOAD QUADROOTS
OLD FILE? YES

*/

1       READ,A,B,C
        PRINT 5,A,B,C
5       FORMAT(2HA=,F9.5,3X,2HB=,F9.5,3X,2HC=F9.5)
        IF(A)3,2,3
2       STOP 2
3       D=B*B-4.*A*C
        IF(D)1,4,4
4       D=SQRT(D)
        A2=A+A
        ROOT1=(-B+D)/A2
        ROOT2=(-B-D)/A2
        PRINT 33, ROOT1, ROOT2
33      FORMAT(/13HREAL ROOTS... ,2E15.6//)
        GO TO 1
        E N D
```

FIGURE 7.7 (Panel C)

Figure 7.8 illustrates the use of various EXEC commands.

```
PLEASE IDENTIFY YOURSELF: 1148111004;SEIDEL

KRONOS III CALLIN TIME IS:

5:41 P.M.  DEC 8, 1970

PHONE LINE 1024
TIMESHARING WILL GO DOWN AT 9:00 TONIGHT. IF YOU HAVE ANY
PROBLEMS BRING YOUR OUTPUT TO ENG 121 OR CALL THE TIMESHARING
OPERATOR AT 1484.

WHAT SYSTEM? EXEC

-NEWS
THE STAFF OF CONTROL DATA CORPORATION BOTH AT THE DOWNEY
SALES OFFICE AND LA JOLLA SYSTEMS DIVISION WISH YOU A
VERY MERRY HOLIDAY SEASON!

-FO

ILLEGAL COMMAND

-FILES
6
PRIME
FIB
PRIMOBJ
FIBOBJ
QUADROOTS
QUADOBJ

-DESCRIBE QUADOBJ

FILE NAME           STATUS          TYPE   RECORDS       UNITS

QUADOBJ             PRIVATE            5         1           2


-DATE
5:43 P.M.  DEC 8, 1970

-STORAGE


MAXIMUM STORAGE ALLOWED   = 250
STORAGE CURRENTLY IN USE = 12

TOTAL CHARGE UNITS THIS MONTH = 21
-MEMORY

FORTRA
012-377-374-000/000-000-000-000/000-000-000-000/000-000-000-000
PROGRAM
013-374-000-000/000-000-000-375/000-000-000-000/000-000-000-000
MEMORY CURRENTLY ASSIGNED IS  8192 WORDS
```

FIGURE 7.8 Various EXEC commands

7.8 Functions in KRONOS III Fortran

The system functions perform frequently used computations. They are called through a replacement statement; they cannot be called with a CALL statement.

Standard mathematical functions.

Name	Function	Type of Argument	Type of Result
ABS (X)	absolute value of *X*	real	real
IABS (I)	absolute value of *I*	integer	integer
SQRT (X)	square root of *X*	real	real
SIN (X)	sine of *X*	real	real
COS (X)	cosine of *X*	real	real
ATAN (X)	principal value of arc tangent of *X*	real	real
ATAN (X,Y)	arc-tangent of *X*/*Y*	real	real
ASIN (X)	principal value of arc sine of *X*	real	real
ACOS (X)	principal value of arc cosine of *X*	real	real
TAN (X)	tangent of *X*	real	real
COT (X)	cotangent of *X*	real	real
ALOG (X)	natural logarithm of *X*	real	real
EXP (X)	e^{X}	real	real
GAMMA (X)	gamma function of *X*	real	real
AINT (X)	integral part of *X*	real	real
FLOAT (I)	convert integer to real	integer	real
IFIX (X)	convert real to integer	real	integer
AMOD (X,Y)	remainder of *X*/*Y*	real	real
MOD (I,J)	remainder of *I*/*J*	integer	integer
SIGN (X,Y)	sign of *Y* times ABS (X)	real	real
ISIGN (I,J)	sign of *J* times IABS (I)	integer	integer
DIM (X,Y)	difference between *X* and the smaller of *X* and *Y*	real	real
IDIM (I,J)	difference between *I* and the smaller of *I* and *J*	integer	integer
ERF (X)	error function of *X*	real	real
ERFC (X)	complement error function of *X*	real	real

Logical functions.

Name	Function	Type of Argument and Result
AND (X,Y)	"logical product" of *X* and *Y*	real
OR (X,Y)	"inclusive or" of *X* and *Y*	real
EOR (X,Y)	"exclusive or" of *X* and *Y*	real
ANOT (X)	"logical complement" of *X*	real
LAND (I,J)	"logical product" of *I* and *J*	integer
LOR (I,J)	"inclusive or" of *I* and *J*	integer
LEOR (I,J)	"exclusive or" of *I* and *J*	integer
NOT (I)	"logical complement" of *I*	integer

Shift functions.

Name	Function	Type of Argument and Result
SHIFT (X,J)	shift *X, J* bits	real
LSHIFT (I,J)	shift *I, J* bits	integer

If *J* is positive, the shifting is left and is circular.

If *J* is negative, the shifting is right, end-off.

MAX and MIN Functions. Eight MAX and MIN functions are provided in the standard Fortran library. They find the maximum or minimum of a variable number of arguments.

Function	Type of Arguments	Type of Result	Value of Result
MINO (i,j, . . .)	integer	integer	minimum of argument list
MIN1 (a,b, . . .)	real	integer	minimum of argument list
MAX0 (i,j, . . .)	integer	integer	maximum of argument list
MAX1 (a,b, . . .)	real	integer	maximum of argument list
AMIN0 (i,j, . . .)	integer	real	minimum of argument list
AMIN1 (a,b, . . .)	real	real	minimum of argument list
AMAX0 (i,j, . . .)	integer	real	maximum of argument list
AMAX1 (a,b, . . .)	real	real	maximum of argument list

RANDOM Number Function. The RANDOM (I) function returns a pseudo-random number, R, where $0 \leqslant R < 1$. There are two modes available: reproducible and non-reproducible. The argument, which may be either real or integer, selects the mode.

If I is negative, the generator is set back to the beginning of its reproducible sequence. The result, R, will be the same as the first time RANDOM is called after loading. If I is zero, the next number in the reproducible sequence is returned as the result. If I is greater than zero, an indefinite number (from 0 to 63) of the next sequential intermediate "values" are skipped before the random number is constructed. The number of values skipped is determined by the least significant portion of the amount of CPU time the individual user has accumulated in the current session.

Exercises

1. Write, compile, and test a Fortran program which reads in 4 integer data values per input case. Calling these variables a, b, c and d, we are required to compute

$$n = a+b+c+d$$

$$f = \frac{n \cdot (\,|a-c| - 1\,)^2}{(a+c) \cdot (b+d)}$$

   Output should be a, b, c, d and f; if $a+b$ is not equal to $c+d$, a warning message should be output along with the computed data. The program must be capable of accepting any number of cases; a, b, c and d are each 3 digits. The program shall terminate whenever a negative or zero value is input for any of the 4 variables.

   While the 4 inputs are integers, note that the calculated value f is a real number. What will be your choice of output format for f?

   Test the program with several sets of data, and check the computer outputs against the results of careful hand calculations.

2. Write, compile, and test a program that will generate and print a list of prime numbers. Format the prime numbers in output lines containing 6 primes each. This problem should include the development of a flowchart as part of the planning stage. The case 2 may be excluded, so that all the desired primes are odd. A single input should be read to determine the upper bound of the list. Avoid using any function references.

3. Write, compile, and test a program to find all the prime factors of any integer representable in the computer available to you. For example, the prime factors of 18 are 2, 3, 3.

4. Write, compile, and test a program to accept 2 dates from a card in the form of 6 digits each (month, day, and year components, each 2 digits). Each date is assumed to be in the twentieth century, between 1901 and 1999; the first date is *later* in the century than the second. Using these restrictions, any year that is a multiple of 4 is a leap year.

   The program shall compute the number of days from the second date to the first one. Output should be double spaced in the form:

   FROM *mm/dd/yy* TO *mm/dd/yy* IS *nnnnn* DAYS.

   Of course, the program should go back to read another data card after producing an output. Be sure to devise adequate test data, such as

   07/04/66 06/30/59
   01/01/60 02/20/59
   01/01/63 12/31/60
   01/01/80 04/04/72
   07/07/67 03/01/64

8

SUBPROGRAMS

8.1 Subprogram Concept

In the original IBM 704 version of Fortran, one "job" had to be written as a single program "unit." For large and complicated jobs, the compilation times became fairly large. When corrections were necessary, the large programs had to be recompiled in their entirety even if only one statement needed to be added, removed, or modified.

User experience quickly showed the need for an improved approach, in which a set of separately compiled program units could be combined as a single job for an execution run.

In the Fortran II language (successor to the original Fortran), this requirement was satisfied by the introduction of the subprogram concept. Separately compiled program units could be combined as one job, and language features were added to enable communication to take place between different program units. (At the same time, small programs would be maintained as single programs).

Subprograms are of two possible kinds: Subroutine and Function. Regardless of kind, a subprogram may be thought of as a self-contained supplement to a Main Program. The interrelationships between a main program and two subprograms may be illustrated by Figure 8.1.

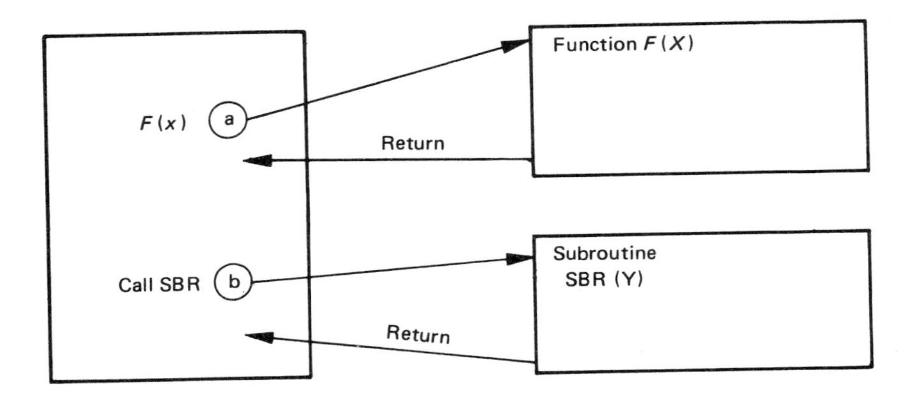

FIGURE 8.1 Subprograms used by a main program

At point (a) in the main program, there is a statement that refers to a function name other than one of the function names in the Library. It is assumed therefore that this function refers to a separately compiled FUNCTION subprogram. In other words, suppose that there appears the statement

```
G = 2.*F(X)-1.0
```

at point (a) in the main program. Since *F*(X) is neither a Library function reference nor a subscripted array variable, the object program will include instructions to call upon a subprogram named F, which must be combined with the main program at execution time. The value of the argument (X) is made available to the subprogram; the function subprogram goes through its procedure to form a resultant value, which is returned to the main program for use as the value of the function reference.

At point (b) in the main program, there is a CALL SBR statement, where SBR is the name of a SUBROUTINE type of subprogram. SBR must be a separately compiled but co-resident subprogram; arguments can be made available to it in a variety of ways, such that upon exit from SBR (returns to main program) the effect of the SBR procedure is reflected in data in the main program.

To summarize matters: a FUNCTION subprogram is entered by another program (or subprogram) by means of a function reference; a SUBROUTINE subprogram is entered by means of a CALL statement.

The following two sections discuss the matter in which subprograms are defined.

8.2 Defining FUNCTION Subprograms

Every FUNCTION subprogram begins with a statement of the form

t FUNCTION name (a_1 ,a_2 , . . . a_n)

where

> *t* is optional and designates type (REAL, INTEGER, DOUBLE PRECISION, COMPLEX, or LOGICAL), and the
>
> a_i are dummy argument variables–at least one dummy argument must be defined.

In the absence of a type modifier for the FUNCTION declaration, the function is assumed type REAL unless the name begins with a letter I through N, in which case type INTEGER is assumed.

Dummy arguments may be mentioned in DIMENSION and/or type declarations.

The function name must be used as a simple (unsubscripted) variable name in the defining subprogram. During every execution of this subprogram, this variable must be assigned a value (normally by virtue of its appearance on the left side of an assignment statement). The value of the variable at the time of executing a RETURN statement in the subprogram becomes the value of the function reference that appeared in the "calling" program. There must be at least one RETURN statement in every FUNCTION subprogram.

A FUNCTION subprogram may not have any of the following statements in it: FUNCTION, SUBROUTINE, BLOCK DATA. (We are fast approaching definitions of the latter two.) Every subprogram has END as its last statement.

8.3 Defining and Calling SUBROUTINE Subprograms

Every SUBROUTINE program begins with a statement of the form

SUBROUTINE name (a_1 , a_2 , . . . ,a_n)

where the dummy argument list is optional.

The name of the subroutine must not appear anywhere else within the subprogram. FUNCTION, SUBROUTINE, and BLOCK DATA statements may not appear elsewhere in the subprogram. Every subprogram has END as its last statement, and must contain at least one RETURN statement in order to revert to the statement after the one that CALLed it.

A subroutine is entered from another program unit (main program or a different subprogram) by means of a CALL statement (in the "other" program unit). The general form of the CALL statement is

CALL name $(a_1, a_2, \ldots a_n)$

The CALL statement must include exactly as many arguments as appear in the separate defining SUBROUTINE statement. Furthermore, the corresponding arguments must agree in terms of type (real, integer, etc.).

A subroutine may assign values to one or more of its arguments in order to "return" results. Furthermore, there is the possibility of producing results in COMMON storage so that a subprogram's effect may be available to other program units; this subject is our next concern.

8.4 Communicating through COMMON

COMMON storage is a data area shareable between different program units of a given job. There are two standard kinds of COMMON storage:

Blank COMMON
Labeled COMMON

plus a third kind unique to the MASTER Operating System for the CDC 3170/3300/3500 series.

Labeled COMMON is intended for the purpose of shared access between a particular set of program units, while blank COMMON is available to all program units.

When a job is being loaded (prepared for execution), the individual program units that are to make up the combined job are surveyed so as to permit efficient allocation of main memory storage space. The first appearance of a particularly named component of COMMON causes the requisite amount of private storage to be set aside for the associated variables. All subsequent program units referencing that same named COMMON component are merely altered to refer to the previously allocated area.

On the other hand, blank COMMON is reserved in such a way as to make all programs refer to the same block of storage, with the largest size specified for it in any program unit being the one that prevails.

The general form of the COMMON statement is:

COMMON $/p_1/d_1/p_2/d_2/ \ldots /p_n/d_n$

where each p_i is a common-region name and each d_i is a list of variable or array names.

Examples:

```
COMMON /CMPLX/Z,ZBAR/INTGR/I,N
COMMON /GREEN/B,B2,BOXES
COMMON // A,P,BETA
```

In the third example, we see an instance of blank COMMON, in that the name (p_1) is null.

A common-region name may be from 1 to 6 alphanumeric characters, of which the first must be alphabetic.

In CDC lower 3000 Fortran, the case of p_i being an integer is permitted in order to reserve "chapter two" common in an additional bank of memory, reserved for use with exceptionally large programs. In this case, $p_i \leqslant 16383$.

8.5 Subprogram Examples

Figure 8.2 illustrates a main program that read coefficients a, b, and c for any quadratic equation, and calls upon the ROOTS subprogram in order to calculate the roots as complex numbers using the quadratic equation.

The following points are significant in examining Figure 8.2:

1. The main program defines five variables in labeled common: X1, X2, A, B, and C. The entire aggregation is assigned the common-region name QUADR.
2. The subprogram (ROOTS) names the constituents of common-region QUADR X1, X2, C1, C2, and C3, respectively. This illustrates the fact that only the common-region name is of any significance on an inter-program basis. However, note that the types of corresponding variables are the same in each program (i.e., X1 and X2 are complex, but the other three are real, regardless of which names are considered).
3. The logic for the execution to stop is based on a blank card following the normal "coefficients" data deck, so that A = 0.0 will signal the end of the job.
4. Compare the syntax of the DATA and COMMON statements; observe that the DATA statement ends with /, while the COMMON statement starts its operands with one.

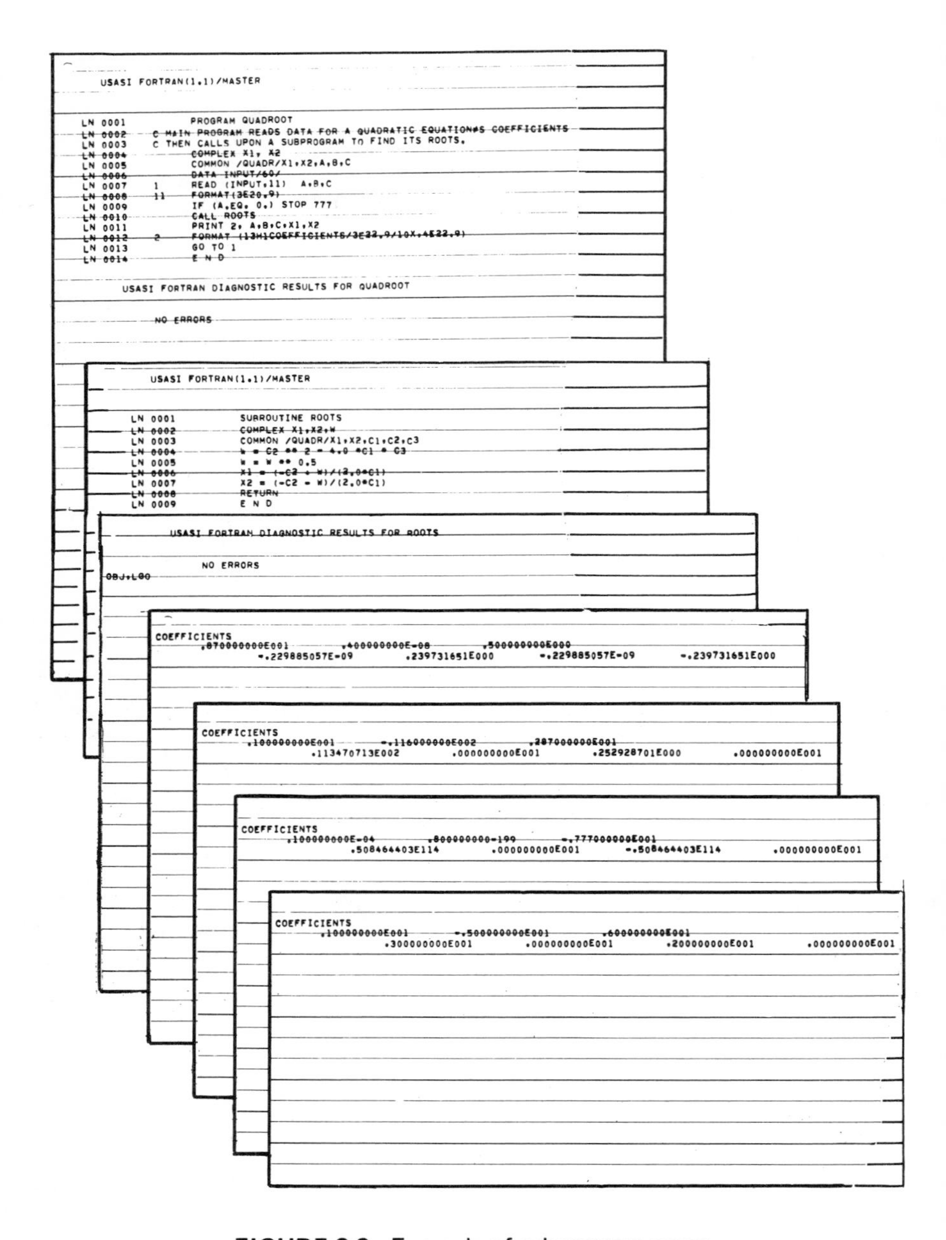

```
USASI FORTRAN(1.1)/MASTER

LN 0001          PROGRAM QUADROOT
LN 0002    C MAIN PROGRAM READS DATA FOR A QUADRATIC EQUATION#S COEFFICIENTS
LN 0003    C THEN CALLS UPON A SUBPROGRAM TO FIND ITS ROOTS.
LN 0004          COMPLEX X1, X2
LN 0005          COMMON /QUADR/X1,X2,A,B,C
LN 0006          DATA INPUT/60/
LN 0007    1     READ (INPUT,11)  A,B,C
LN 0008    11    FORMAT(3E20.9)
LN 0009          IF (A.EQ. 0.) STOP 777
LN 0010          CALL ROOTS
LN 0011          PRINT 2, A,B,C,X1,X2
LN 0012    2     FORMAT (13H1COEFFICIENTS/3E22.9/10X,4E22.9)
LN 0013          GO TO 1
LN 0014          E N D

   USASI FORTRAN DIAGNOSTIC RESULTS FOR QUADROOT

      NO ERRORS
```

```
USASI FORTRAN(1.1)/MASTER

LN 0001          SUBROUTINE ROOTS
LN 0002          COMPLEX X1,X2,W
LN 0003          COMMON /QUADR/X1,X2,C1,C2,C3
LN 0004          W = C2 ** 2 - 4.0 *C1 * C3
LN 0005          W = W ** 0.5
LN 0006          X1 = (-C2 + W)/(2.0*C1)
LN 0007          X2 = (-C2 - W)/(2.0*C1)
LN 0008          RETURN
LN 0009          E N D
```

```
   USASI FORTRAN DIAGNOSTIC RESULTS FOR ROOTS

      NO ERRORS
OBJ.L00
```

```
COEFFICIENTS
     .870000000E001      .400000000E-08      .500000000E000
          -.229885057E-09      .239731651E000      -.229885057E-09      -.239731651E000
```

```
COEFFICIENTS
     .100000000E001      -.116000000E002      .287000000E001
          .113470713E002      .000000000E001      .252928701E000      .000000000E001
```

```
COEFFICIENTS
     .100000000E-04      .800000000-199      -.777000000E001
          .508464403E114      .000000000E001      -.508464403E114      .000000000E001
```

```
COEFFICIENTS
     .100000000E001      -.500000000E001      .600000000E001
          .300000000E001      .000000000E001      .200000000E001      .000000000E001
```

FIGURE 8.2 Example of subprogram usage

5. In ROOTS, the discriminant

   $$b^2 - 4ac$$

   is computed as a separate complex entity, whose square root is then taken by exponentiation, since the real SQRT function cannot yield a complex result.
6. The Fortran compiler accepts multiple program units in source form, so that the main program and the ROOTS subprogram can be considered a single source input (i.e., the Fortran compiler is only "called" once by job control cards that precede the set of source program inputs).

8.6 Adjustable Dimensions

In any subprogram, an array may be defined as having variable dimensions, using an integer variable as the dimensions, as in the following example.

```
DIMENSION ARRAY (I)
```

In this case, I must be a dummy argument. The value of I in effect at the time of entering the subprogram determines the size of the array during that particular execution of the subprogram. The user must take care that the value of I is not negative or zero.

8.7 BLOCK DATA Subprograms

The BLOCK DATA subprogram is used to place data in variables located in labeled common storage. The subprogram may contain only the following kinds of statements:

```
BLOCK DATA
DATA
COMMON
DIMENSION
type (REAL, INTEGER, COMPLEX, etc.)
END
```

Executable statements are not allowed in the subprogram. Blocks of common storage and the variables therein may be specified and data entered into the variables of the common blocks.

The following rules pertain to BLOCK DATA subprograms:

1. The first statement of the program must be the BLOCK DATA statement. There are no other components (words or operands) in the statement.
2. The COMMON statements in the subprogram must include all the elements of the common blocks, although not all of the elements need be specified in the DATA statement.
3. The DATA statement must define only elements listed in the COMMON statements.
4. The last statement of the subprogram must be an END statement. For example,

```
BLOCK DATA
DIMENSION TABLE (4)
COMMON  /AA/X/BB/TABLE
DATA    X/5.06/TABLE/2*0.0,-1.7,5.E-4
END
```

8.8 EXTERNAL Statement

An EXTERNAL statement is of the form

EXTERNAL $v_1, v_2, \ldots v_n$

where each v_i is an external procedure name.

Appearance of a name in an EXTERNAL statement declares that name to be an external procedure name, which may be used in the argument list passed to a subprogram.

Example:

```
EXTERNAL SIN, SQRT
          .
          .
          .
CALL OTHER (A,B,SIN)
          .
          .
          .
CALL OTHER (E2,E1,SQRT)
```

Exercises

1. Write, compile, and test a FUNCTION subprogram called CROSS which accepts as input the following arguments:

   (a) N, the number of real values in each of two one-dimension arrays called A and B.

   (b) M, a value representing the number of terms by which series B is "shifted," so that the cumulative cross product

   $$\text{CROSS} = \sum_{i=1}^{n-m} A_{i+m} \cdot B_i$$

   may be computed.

   (c) A and B.

   In the above notation, n is the value given for dummy argument N and m is the value of dummy argument M. The program should check that $n-m \geqslant 1$; otherwise, CROSS should be assigned the value 0.

   The program solution to this problem should use adjustable dimensions and one DO statement.

2. Write, compile, and test a program that will read in any list of numbers, print them out, call a subroutine to sort the numbers into ascending order, and print out the results. Use the adjustable dimensions feature in the subprogram, which should be entered from the main program by the following statement:

   ```
   CALL SORTER (N, I)
   ```

   where I is the one-dimensional array of integers to be sorted, and N is the number of integers that were read into the array. The value of N may not exceed 100. Devise test data that contains an unorderly mixture of up to 100 numbers, some zero, some negative, some positive, and mostly (but not necessarily all) unique.

   The program should recycle for a new set of inputs. Each set can be of any length, from 1 to 100, and shall be preceded by its own set count on a separate input card.

3. You are hired by the Pollutant Motor Co. to program their payroll application.

Card Input	Columns
1. Man Number	1–5
2. Department (integer 1–9)	7
3. Hours worked (integer)	9–10
4. Pay Rate (real, 2 places for cents)	12–16 e.g., 05.35

Printed Output
1. Man Number
2. Department
3. Hours worked
4. Regular hours
5. Regular pay rate/hr.
6. Regular pay
7. Overtime hours
8. Overtime pay (at 1.5 times regular rate)
9. Total pay

Be sure to leave a few spaces between each output field.

Use at least 15 data cards, and use a variety of data values to test out each contingency (e.g., hours over 40, rates that are not integers, etc.).

4. You are to augment the program developed in problem 3 in the following ways:
   - (a) Print a report title, then skip 2 lines.
   - (b) Print column headings, followed by 1 skip line.
   - (c) Print totals for each department number.
   - (d) Print company totals.
   - (e) All total lines should be double spaced, and preceded by an appropriate line of descriptive text.

9

ADVANCED INPUT-OUTPUT

9.1 BCD Input-Output

BCD (binary coded decimal) refers to input or output in character form, converted under control of a FORMAT statement. BCD is the form in which cards are read or punched, or printed data is developed.

Within the computer, data is represented in non-BCD form, usually binary. In the CDC 3170/3300/3500 computers, data is represented by one or more binary "words", each 24 binary digits (bits) in length. This internal format is required in order for the computer to perform any arithmetic in an efficient manner.

On input (READ statement) using a FORMAT statement, data in BCD form (e.g., 7.63E+04) is scanned and converted to the appropriate binary form (floating-point, in the case of the real number 7.63E+04). The associated format specification (say, F8.0) specifies that the scanning is over a field of 8 characters in the input card, and the floating-point value is stored in a real variable (two computer words).

On output (PRINT, PUNCH, or WRITE statement) using a FORMAT statement, data is converted from internal format to character form, as directed by a matching format specification. The number of characters output is given by the field width of the format specification.

9.2 Binary Input-Output

Binary output is a highly efficient form of output normally employed when preparing data for magnetic tape or disk files. Data in binary format is recorded for future use by the computer; for example, a magnetic tape may be used as temporary external storage during a Fortran program run, or may even be passed from one step of a job to another, when several separate Fortran programs run serially in a production system.

Binary input-output data is not converted, since it is already in the desired internal form. Consequently, no format statement reference appears in the READ and WRITE statements that govern this type of data transmission.

Binary output is created by the unformatted Write statement:

WRITE (*i*) list

Binary input is created by the unformatted Read statement:

READ (*i*) list

In each of the above statement, *i* is a unit number, either an integer constant or a simple integer variable.

In the binary Write statement, one "logical" binary record is written. The number of variables in the list determines the number of physical records. Each physical record contains 128 words: the first word is a flag field; the remaining 127 words contain transmitted data. When the amount of data to be transmitted exceeds 127 computer words, enough physical records are created in order to complete the requested Write action.

The flag word contents indicate to the Fortran I/O package how many consecutive physical records constitute a single logical record. A logical record consists of all consecutive physical records having a zero flag word (if any), plus the first physical record having a non-zero flag word; the value in the latter flag word is k, equal to the number of physical records into which the logical record was broken.

In the binary Read statement, one logical record is read into storage locations named by the listed variables. The record must have been written in binary mode by an unformatted Write statement. Flag word(s) are not transmitted into storage by the Read statement.

The number of data words required by the input list must not exceed the number of data words that constitute the logical record. If there is no list in a binary read statement, the statement bypasses one entire logical record.

9.3 File Control Statements

Statements are available for manipulating files other than the standard system input (unit 60), output listing (unit 61), and punch (unit 62).

Auxiliary statements to rewind a magnetic tape file, backspace over one logical record in a file, or write an end-of-file mark have the following formats:

```
REWIND      i
BACKSPACE   i
ENDFILE     i
```

9.4 Formats Stored in Arrays

BCD reads and writes may use an array name as the format reference, instead of a statement number. In this case, the array must contain the format in Hollerith form.

Use of this facility permits the data formats to be varied at execution time, without recompilation of the program. In this case, input consists of the format information (parenthesized) as well as the data values, which must of course be prepared in conformance with the new format.

Example:

```
      DIMENSION FMT (10)
      READ (60,11) FMT
11    FORMAT (10A8)
      READ (60, FMT) X, Y, Z
           .
           .
           .
```

Data Card No. 1

(E20.10, F14.8 6X, F20.7)

Data Card No. 2

columns 1–14: 375.4939972E-1

columns 25–34: 7691021218

columns 41–43: 8.7

9.5 End-of-File Detection

In the CDC lower 3000 Fortran, the function IFEØF (*i*) is provided in order to determine whether unit *i* reached the end-of-file point on the last attempted Read statement.

The value of the function reference is zero unless the end-of-file point was reached in the last read, in which case the function value is -1.

In Fortran systems where this function is not available, it is common practice to place a special data card distinct from any proper data. A blank card often suffices for this purpose, as shown in the following example.

```
      READ (60, 4)I, A, S
   4  FORMAT (I5, 2F12.5)
      IF (I) 5, 5, 10
   5  STOP 717
  10  PRINT 11, I, A, S
  11  FORMAT (1H1,9HCASE NO., I5/3H A=,F12.5,2X,2HS,F12.5)
       .
       .
       .
```

Some Fortran systems, such as those in the IBM System 360, XDS Sigma, and RCA Spectra 70 families, provide a facility for detection of the end-of-file condition built into a READ statement. The applicable statement forms are

READ (*i*, *f*, EØF = *s*) list

for formatted (BCD) reading from unit *i*, and

READ (*i*, EØF = *s*) list

for unformatted (binary) reading from unit *i*. In each case, *s* is a statement number to which control will pass if the attempt to read encounters an end-of-file condition.

IBM System 360/67

Xerox Data Systems Sigma 9

9.6 NAMELIST Input-Output

The NAMELIST statement, available in IBM System 360 and 370 series Fortran, permits the user to have self-identified input and output.

A basic form of the NAMELIST statement is

NAMELIST / name_1 / list_1 / name_2 / list_2 . . .

where name is unique with respect to all other variables in the program unit (except for dummy arguments in any AFDS), and list is one or more variable names, unsubscripted. Successive variable names in a list are separated by commas.

The effect of a NAMELIST statement is to encompass an indicated list of simple variables or arrays by a single group name. The group name is used in a special form of READ or WRITE, which controls the transmission of values between variables in the main computer memory and external media, by associating the name and value.

A NAMELIST-type read statement has the general form

READ (unit, name)

where name is a "group" name designated in a previous NAMELIST statement, and unit specifies a logical unit number.

Input data prepared for NAMELIST type reading resembles replacement statements separated by commas. That is, both the variable name and value are on the input cards. The following are IBM NAMELIST input rules.

1. The first character of every card is ignored.

2. The first card to be read must have an ampersand (&) in column 2, followed by the group name specified in the READ statement.

3. The last card to be read for each namelist READ must contain &END in columns 2–5.

Example:

```
        DIMENSION A(3)
        NAMELIST / INFO / A, I, B, P
        A(1)   = 0.0
        A(2)   = 1.5
        A(3)   = -3.2
        INPUT = 60
            .
            .
            .
     75 READ (INPUT, INFO)
```

Data Card No. 1

```
&INFO
```

Data Card No. 2

```
A(2) = 5.601, I = -7, B = 1.0E-9
```

Data Card No. 3

```
&END
```

In this example, note that statement 75 results in assignment of the value 5.601 to the second member of the real array A; I is assigned the value -7, and B assumes the value 1.0E-9. Each of the variables A, I, and B is designated as belonging to Namelist group "INFO." However, in this case, not all possible value replacements that could be made by the input are done so in this illustration. Specifically, note that execution of statement 75 does not alter the values of A(1), A(3), or P. The same NAMELIST-type READ statement, executed at a later time, could assign new values to any of the variables A, I, B, or P, depending on the content of the next set of input cards.

The form of WRITE used with NAMELIST groups is

WRITE (unit, name)

The rules governing this form of output are such that the results of one execution of the WRITE may themselves be input by one execution of the corresponding NAMELIST-type READ statement.

Since only simple (unsubscripted) variables or array names may appear in a NAMELIST statement, it is not possible to output array elements individually on a "selective" basis. Output of an array in its entirety is in the form

array = value, value, etc. . .

This form of pseudo-replacement statement is valid as input to an array. For arrays of more than one dimension, values are ordered corresponding to variation of the leftmost subscript most rapidly. (For a 2-dimensional array, this has the effect of specifying values in columnar order.)

9.7 PUNCH Statement

A statement of the form

PUNCH n, list

causes BCD output developed according to format number *n* to be punched onto 80-column cards.

10

COMPARATIVE ASPECTS OF FORTRAN

10.1 Standardization

Fortran has been standardized by the American National Standards Institute (formerly called United States of America Standards Institute, after an earlier period when the name was American Standards Association).

The standard for Fortran took two forms: Basic Fortran (often called Fortran II) and Fortran (often called Fortran IV).

In compliance with this standardization effort, most manufacturers supply one or more versions of Fortran with their computer systems.

At the same time, it must be realized that *compatibility* is often incomplete, although the languages on two different computers may appear to be identical. But if we look beneath the surface, we often encounter quite a few significant differences.

10.2 Some Differences in Language

Some Fortran compilers permit the names of variables to be of lengths in excess of 6 characters. In some cases, a name can be up to 8 characters long; in others, names can be of arbitrary length, but only the first 6 or 8 characters are used to differentiate between the various names. In fact, the CDC 3170/3300/3500 systems accept 8-character names, but in this book we have stressed the standard maximum size of 6.

There are many Fortran compilers which have additional features to enhance their attractiveness in the software marketplace. As an example, character variables are sometimes provided for a Fortran compiler; others provide ENCODE and DECODE statements which permit formatted data conversion between variables in storage (rather than between external media and storage); some compilers support double precision complex data; many other special features permit the user to specify alternative variable naming or size allocation conventions at compilation time.

10.3 Differences in Implementation

It must be remembered that both real and integer data representations are machine-dependent, so that the degree of accuracy obtained in calculations is affected by the particular hardware (and software, too, sometimes) that a program is run on.

Not every compiler produces the same result from a given statement. Consider the suspect program in Figure 10.1. What output do you expect the program to produce?

```
1  FORMAT (1H, I5)
   I = 3
   PRINT 1, (I, I = I, I, I)
   STOP 7
   END
```

FIGURE 10.1 Tricky DO-implied list

This program has been executed on the XDS Sigma 7, CDC 6600, GE 415, IBM 360/75, Univac 1108, and CDC 3170 computers. The output was one line containing the value 3, in all instances except two. Output from the IBM 360/75, using the OS/360 Fortran H compiler, was the series of numbers 3, 6, 12, 24, 48, 96, 192, 384, etc., so that execution terminated only when the time limit was exceeded (although field capacity was rather rapidly exceeded, and not all output was proper). The other computer that produced this result was an XDS Sigma 7, used through a time-sharing terminal.

Some compilers are famous for handling the DO loop differently; for example, if the initial values of J and K are 1 and 0, respectively, the statement DO 8 I = J, K would (in these non-standard implementations) not pass control into the DO range, since the end test (I > K?) is performed *prior* to entering into the DO range.

Finally, some computers use different punched card codes for certain characters. The CDC lower 3000 computers employ the Hollerith (BCD) system[13]; IBM System 360 computers employ the extended binary-coded-decimal interchange code (EBCDIC), so that the punches in a card for the symbols

( ) + ' =

must be different in these two different systems. BCD is the usual code for computers that represent a character in internal computer memory as a 6-bit entity (called a *byte*); EBCDIC is used in many third generation computer systems that represent a character in memory as an 8-bit byte.

13. Refer to Appendix A.

ANSWERS TO SELECTED EXERCISES

Exercise 2, Chapter 2

FORTRAN Coding Form

PROGRAM Exercise 2, Chapter 2
PROGRAMMER
DATE
PUNCHING INSTRUCTIONS
0 = ZERØ 1 = ØNE 2 = TWO
Ø = ALPHA O I = ALPHA I Ƶ = ALPHA Z
PAGE ____ OF ____

```
C     EXERCISE 2, CHAPTER 2...INTEGER DIVISIØN RESEARCH.
      INEG= -2
      IPØS= 2
      JNEG= -3
      JPØS= 3
      K1 = INEG/JPØS
      K2 = IPØS/JNEG
      PRINT 90, K1, K2
90    FØRMAT (2I10)
      INEG= -3
      IPØS= 3
      JNEG= -2
      JPØS= 2
      K1=INEG/JPØS
      K2=IPØS/JNEG
      PRINT 90, K1, K2
      STØP
      END
```

Exercise 3, Chapter 2

FORTRAN Coding Form

PROGRAM: Exercise 3, Chapter 2 — PROGRAMMER: — DATE: — PUNCHING INSTRUCTIONS: O = ZERØ, 1 = ØNE, 2 = TWO; Ø = ALPHA O, I = ALPHA I, Ƶ = ALPHA Z — PAGE ___ OF ___

```
C     EXERCISE 3, CHAPTER 2...FIBØNACCI SERIES.
      K1=0
      K2=1
5     K3=K2+K1
      R2=K2
      R3=K3
C NØW CALCULATE RATIØ AS A REAL RATIØ.
      RATIØ=R3 / R2
      PRINT 1, K3, RATIØ
C NØW MØVE K2 TØ K1, K3 TØ K2 FØR REPEAT ØF ABØVE PRØCESS.
      K1 = K2
      K2 = K3
      GØ TØ 5
C THIS IS A GREAT PRØGRAM, BUT IT IS NØT FINISHED...
C HØW DØES IT TERMINATE ??  THE ABØVE 'GØ TØ 5' MUST BE REPLACED BY
C SØME TYPE ØF TEST AND CØNDITIØNAL BRANCH. CØNSIDER 2 ALTERNATIVES.
C     1. INTRØDUCE A VARIABLE, INITIALLY ZERØ, AND CØUNT THE NØ. ØF
C        TERMS, AND MAKE A TEST TØ STØP WHEN THIS CØUNT ATTAINS SØME
C        MAXIMUM.
C     2. EVENTUALLY, ADDING 2 INTEGERS PRØDUCES A SUM THAT WILL NØT FIT
C        IN THE SPACE RESERVED BY THE CØMPILER, AND 'ØVERFLØW' CAUSES K3
C        TØ GØ AWRY, SØ THAT A TEST CAN BE MADE FØR K3 LESS THAN K2.
1     FØRMAT ( I15, F15.10)
      END
```

Exercise 1, Chapter 3

FORTRAN Coding Form

PROGRAM: Exercise 1, Chapter 3

PROGRAMMER: DATE:

PUNCHING INSTRUCTIONS: 0 = ZERØ, 1 = ØNE, 2 = TWO; Ø = ALPHA O, I = ALPHA I, Z = ALPHA Z

PAGE ____ OF ____

```
C     EXERCISE 1, CHAPTER 3...WALLIS' EQUATIØN RØØTS BY BRACKETING.
      X = 3.0
      DX = 0.1
      Y = X**3 -2.0*X -5.0
4     XZ = X - DX
      Z = XZ**3 -2.0*XZ-5.0
      IF (Z*Y) 10, 5, 5
5     X=XZ
      Y=Z
      GØ TØ 4
10    PRINT 13, X, Y, DX
13    FØRMAT (3E20.10)
      DX = DX*0.1
      IF (DX-1.0E-11) 99,99,4
99    STØP 3
      END
```

Exercise 1, Chapter 5

FORTRAN Coding Form

PROGRAM *Exercise 1, Chapter 5* | PUNCHING INSTRUCTIONS: O = ZERØ, 1 = ØNE, 2 = TWO, Ø = ALPHA O, I = ALPHA I, Z = ALPHA Z | PAGE ____ OF ____

PROGRAMMER | DATE

TYPE STATEMENT NUMBER | CONT. | FORTRAN STATEMENT | IDENTIFICATION SEQUENCE

1 2 3 4 5 6 7 8 9 10 11 12 13 14 15 16 17 18 19 20 21 22 23 24 25 26 27 28 29 30 31 32 33 34 35 36 37 38 39 40 41 42 43 44 45 46 47 48 49 50 51 52 53 54 55 56 57 58 59 60 61 62 63 64 65 66 67 68 69 70 71 72 73 74 75 76 77 78 79 80

```
C     EXERCISE 1, CHAPTER 5...REVERSING 72-MEMBER INTEGER ARRAY.
      DIMENSIØN I (72)
72    FØRMAT (12I6)
      READ 72, I
      PRINT 73, I
73    FØRMAT (1H1/12(6I12))
      DØ 17 J=1, 36
      KTEMP = I (-1*J+73)
      I (-1*J+73) = I (J)
17    I (J) = KTEMP
      PRINT 73, I
      STØP 17
      END
```

1 2 3 4 5 6 7 8 9 10 11 12 13 14 15 16 17 18 19 20 21 22 23 24 25 26 27 28 29 30 31 32 33 34 35 36 37 38 39 40 41 42 43 44 45 46 47 48 49 50 51 52 53 54 55 56 57 58 59 60 61 62 63 64 65 66 67 68 69 70 71 72 73 74 75 76 77 78 79 80

Exercise 4, Chapter 5 (Monkey–Coconut Problem)

Mathematical Approach

Original Size

$t = 4n + 1$ .... representing first apportionment
$3n = 4p + 1$ .... representing second apportionment
$3p = 4q + 1$ .... representing third apportionment
$3q = 4r + 1$ .... representing fourth apportionment
$3r = 4s + 1$ .... representing the group's final action

Approach:

Let $s = 1$ and evaluate the last expression $(4s + 1)$; test this quantity to see if it is really a multiple of 3. If so, proceed to substitute the value of r in the above equation; continue upward through the equations as long as integers that are a multiple of 3 can be obtained.

Whenever an untenable situation obtains, s is increased to the next trial value and the entire process is repeated, until an answer is derived for the entire system.

Programming Approach

Restate the previous equations in subscripted fashion; where

s becomes k_1

r becomes k_2

q becomes k_3

p becomes k_4

$l_1 = 4k_1 + 1 = 3k_2$

$l_2 = 4k_2 + 1 = 3k_3$

$l_3 = 4k_3 + 1 = 3k_4$

$l_4 = 4k_4 + 1 = 3k_5$

$t = 4k_5 + 1$

Exercise 4, Chapter 5 (Cont'd)

Flowchart

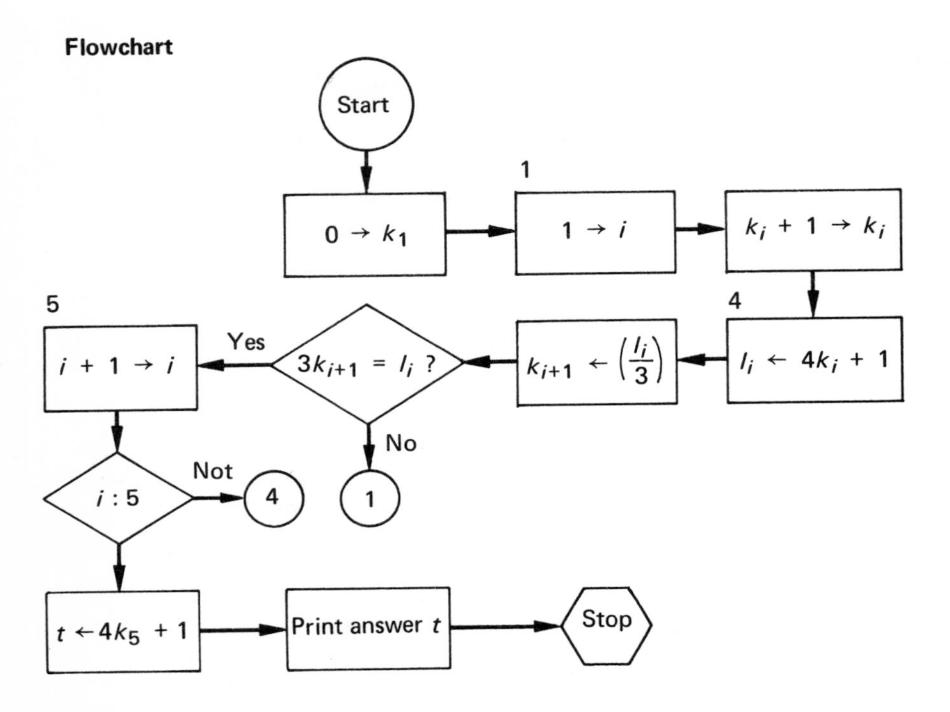

Fortran Implementation

```
      DIMENSIØN  K (5), L (4)
      K (1) = 0
1     I = 1
      K (I) = K (I) + 1
4     L (I) = 4 * K (I) + 1
      K (I + 1) = L (I) / 3
      IF (3 * K (I + 1) - L (I) )  1, 5, 6
6     PRINT, "COMPUTING ERRØR"
5     I = I + 1
      IF (I - 5) 4, 9, 6
9     J = 4 * K (5) + 1
      PRINT 7, J, K, L
7     FØRMAT ( I8 // 5I8 // 6I8)
      STØP 33
      E N D
```

Answers:	J = 1021	("t")				
	K array:	80	107	143	191	255
	L array:	321	429	573	765	

Exercise 2, Chapter 7 (Prime Numbers Generation Program Run on Univac 1108

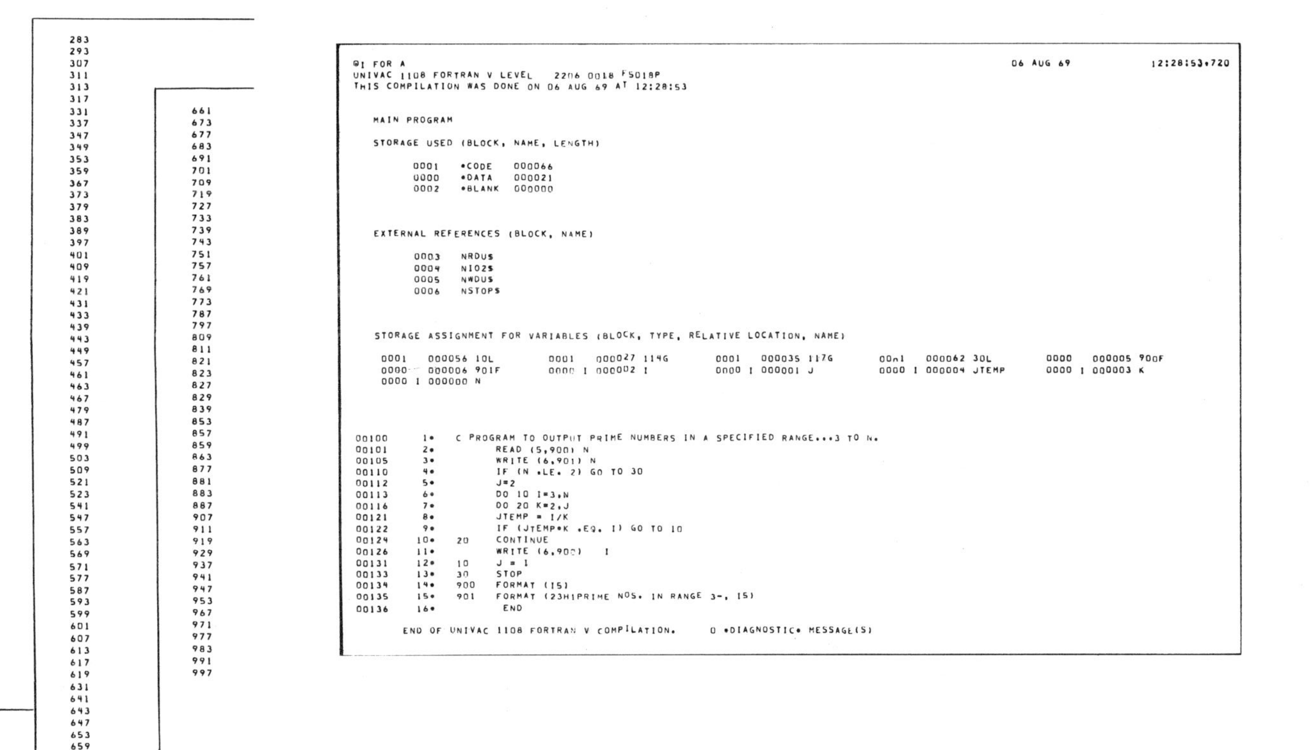

```
PRIME NOS. IN RANGE 3- 1000
    3
    5
    7
   11
   13
   17
   19
   23
   29
   31
   37
   41
   43
   47
   53
   59
   61
   67
   71
   73
   79
   83
   89
   97
  101
  103
  107
  109
  113
  127
  131
  137
  139
  149
  151
  157
  163
  167
  173
  179
  181
  191
  193
  197
  199
  211
  223
  227
  229
  233
  239
  241
  251
  257
  263
  269
  271
  277
  281
  283
  293
  307
  311
  313
  317
  331
  337
  347
  349
  353
  359
  367
  373
  379
  383
  389
  397
  401
  409
  419
  421
  431
  433
  439
  443
  449
  457
  461
  463
  467
  479
  487
  491
  499
  503
  509
  521
  523
  541
  547
  557
  563
  569
  571
  577
  587
  593
  599
  601
  607
  613
  617
  619
  631
  641
  643
  647
  653
  659
  661
  673
  677
  683
  691
  701
  709
  719
  727
  733
  739
  743
  751
  757
  761
  769
  773
  787
  797
  809
  811
  821
  823
  827
  829
  839
  853
  857
  859
  863
  877
  881
  883
  887
  907
  911
  919
  929
  937
  941
  947
  953
  967
  971
  977
  983
  991
  997
```

```
@I FOR A                                                                    06 AUG 69        12:28:53.720
UNIVAC 1108 FORTRAN V LEVEL   2206 0018 FS018P
THIS COMPILATION WAS DONE ON 06 AUG 69 AT 12:28:53

  MAIN PROGRAM

  STORAGE USED (BLOCK, NAME, LENGTH)

       0001    •CODE    000066
       0000    •DATA    000021
       0002    •BLANK   000000

  EXTERNAL REFERENCES (BLOCK, NAME)

       0003    NRDU$
       0004    NIO2$
       0005    NWDU$
       0006    NSTOP$

  STORAGE ASSIGNMENT FOR VARIABLES (BLOCK, TYPE, RELATIVE LOCATION, NAME)

   0001    000056 10L       0001    000027 114G     0001    000035 117G     0001    000062 30L      0000    000005 900F
   0000    000006 901F      0000 I  000002 I        0000 I  000001 J        0000 I  000004 JTEMP    0000 I  000003 K
   0000 I  000000 N

00100      1*    C PROGRAM TO OUTPUT PRIME NUMBERS IN A SPECIFIED RANGE...3 TO N.
00101      2*          READ (5,900) N
00105      3*          WRITE (6,901) N
00110      4*          IF (N .LE. 2) GO TO 30
00112      5*          J=2
00113      6*          DO 10 I=3,N
00116      7*          DO 20 K=2,J
00121      8*          JTEMP = I/K
00122      9*          IF (JTEMP*K .EQ. I) GO TO 10
00124     10*    20    CONTINUE
00126     11*          WRITE (6,900)   I
00131     12*    10    J = I
00133     13*    30    STOP
00134     14*    900   FORMAT (I5)
00135     15*    901   FORMAT (23H1PRIME NOS. IN RANGE 3-, I5)
00136     16*           END

     END OF UNIVAC 1108 FORTRAN V COMPILATION.      0 *DIAGNOSTIC* MESSAGE(S)
```

Exercise 4, Chapter 7

```
      USASI FORTRAN(1.1)/MASTER                                              12/12/70

 LN 0001    C    DIFFERENCE IN DAYS BETWEEN 2 DATES  (MM/DD/YY., 20TH CENTURY ONLY.
 LN 0002    C    FIRST DATE MUST BE THE LARGER.
 LN 0003         DIMENSION MO(12), M(2), ID(2), IY(2), N(2)
 LN 0004         MO (1) = 0
 LN 0005         MO (2) = 31
 LN 0006         MO (3) = 59
 LN 0007         MO (4) = 90
 LN 0008         MO (5) = 120
 LN 0009         MO (6) = 151
 LN 0010         MO (7) = 181
 LN 0011         MO (8) = 212
 LN 0012         MO (9) = 243
 LN 0013         MO(10) = 273
 LN 0014         MO(11) = 304
 LN 0015         MO(12) = 334
 LN 0016    1    READ 901, M(1), ID(1), IY(1), M(2), ID(2), IY(2)
 LN 0017         IF (M(1))99,99,2
 LN 0018    2    PRINT 902, (M(I), ID (I), IY(I), I = 1, 2 )
 LN 0019    C    DAY-NUMBER-IN-YEAR... J IS LEAP YEAR ADJUSTMENT.
 LN 0020         DO 10 I = 1,2
 LN 0021         J = 0
 LN 0022         K = M(I)
 LN 0023         IF (K-2) 8,8,5
 LN 0024    5    IF ( (IY(I)/4)*4 - IY(I) ) 8,6,8
 LN 0025    6    J = 1
 LN 0026    8    N (I) = ID(I) + J + MO(K)
 LN 0027   10    CONTINUE
 LN 0028    C    COMPUTE K = YEAR DIFFERENCE
 LN 0029         K = 0
 LN 0030         GO TO 27
 LN 0031   20    IF ( (IY(2)/4) * 4 - IY(2) ) 22,21,22
 LN 0032   21    K = K + 1
 LN 0033   22    K = K + 365
 LN 0034         IY (2) = IY(2) + 1
 LN 0035   27    IF (IY(1) - IY(2))  1,25,20
 LN 0036   25    K = K + N(1) - N(2)
 LN 0037         PRINT 903, K
 LN 0038         GO TO 1


      USASI FORTRAN(1.1)/MASTER                                              12/12/70

 LN 0039   99    CALL EXIT
 LN 0040  901    FORMAT (3I2, 1X, 3I2 )
 LN 0041  902    FORMAT (1H0, I2, 1H/,I2, 1H/, I2, 4H TO , I2, 1H/, I2, 1H/, I2 )
 LN 0042  903    FORMAT (21X, I6)
 LN 0043         END

         USASI FORTRAN DIAGNOSTIC RESULTS FOR FTN.MAIN

                NO ERRORS
OBJ.LGO


12/30/68 TO  1/ 1/67
                     729
12/30/69 TO  1/ 1/67
                    1094
 2/29/68 TO  2/25/63
                    1830
 7/ 4/69 TO  6/30/59
                    3657
 3/ 1/58 TO  3/ 1/44
                    5113
 3/30/50 TO  2/ 1/46
                    1518
 7/ 1/69 TO  7/ 1/68
                     365
12/31/75 TO 12/31/60
                    5478
```

Exercise 1, Chapter 8

FORTRAN Coding Form

PROGRAM	Exercise 1, Chapter 8	PUNCHING INSTRUCTIONS	O = ZERØ	1 = ØNE	2 = TWO	PAGE ____ OF ____
PROGRAMMER	DATE		Ø = ALPHA O	I = ALPHA I	Ƶ = ALPHA Z	

STATEMENT NUMBER	CONT.	FORTRAN STATEMENT	IDENTIFICATION SEQUENCE
		FUNCTIØN CRØSS(N,M,A,B)	
		DIMENSIØN A(N), B(N)	
		CRØSS = 0.0	
		IF ((N-M) .LT. 1) GØ TØ 909	
		IMAX = N-M	
		DØ 900 I = 1, IMAX	
		IM = I+M	
900		CRØSS = CRØSS + A(IM)+B(I)	
909		RETURN	
		END	

APPENDICES

Appendix A BCD character representations

Character	Card Punches	Internal BCD Code	Character	Card Punches	Internal BCD Code
A	12–1	21	X	0–7	67
B	12–2	22	Y	0–8	70
C	12–3	23	Z	0–9	71
D	12–4	24	0	0	00
E	12–5	25	1	1	01
F	12–6	26	2	2	02
G	12–7	27	3	3	03
H	12–8	30	4	4	04
I	12–9	31	5	5	05
J	11–1	41	6	6	06
K	11–2	42	7	7	07
L	11–3	43	8	8	10
M	11–4	44	9	9	11
N	11–5	45	blank	none	60
O	11–6	46	=	3–8	13
P	11–7	47	+	12	20
Q	11–8	50	–	11	40
R	11–9	51	*	11–4–8	54
S	0–2	62	/	0–1	61
T	0–3	63	(	0–4–8	74
U	0–4	64	)	12–4–8	34
V	0–5	65	.	12–3–8	33
W	0–6	66	,	0–3–8	73

Appendix B Typical compilation diagnostic messages

Code	Message Text	Further Explanation
1004	FUNCTION STATEMENT FORMAT ERROR	Can be caused by no arguments, too many commas, illegal operator, unmatched parentheses, or no function name
1005	ILLEGAL FORMAL PARAMETER	Illegal dummy argument in SUBROUTINE or FUNCTION statement
1006	FORMAT ERROR IN EXTERNAL STATEMENT	Can be caused by too many commas, no list, illegal operator
1007	ILLEGAL USE OF NAME IN EXTERNAL STATEMENT	Name has appeared in COMMON, DIMENSION, TABLE or EQUIVALENCE statement
1008	EQUIVALENCE FORMAT ERROR	Caused by unmatched parentheses, too many commas, or illegal operator
1009	CONSTANT .GT. 2**15-1 IN DECLARATIVE STATEMENT	Can be caused by all type statements, DIMENSION, and COMMON statements
1010	ILLEGAL OPERATOR IN DECLARATIVE STATEMENT	Can be caused by DIMENSION, COMMON, and all type statements
1011	TOO MANY COMMAS OR NO LIST IN DECLARATIVE STATEMENT	Can be caused by DIMENSION, COMMON, and all type statements
1012	MISSING) IN DECLARATIVE STATEMENT	Can be caused by DIMENSION, COMMON, and all type statements
1013	MISSING , IN DECLARATIVE STATEMENT	Can be caused by DIMENSION, COMMON, and all type statements
1014	DIMENSIONING FORMAT ERROR	Can be caused by DIMENSION, COMMON, and all type statements
1015	NAME TYPED MORE THAN ONCE	Name has appeared in type statements more than one time. First type will be used if execution is forced
1016	ILLEGAL USE OF NAME OR NAME DIMENSIONED MORE THAN ONCE IN DECLARATIVE STATEMENT	Can be caused by DIMENSION, COMMON, and all type statements
1017	ILLEGAL USE OF SUBSCRIPT	Can be caused by DIMENSION, COMMON, and all type statements

Appendix B–*Continued*

Code	Message Text	Further Explanation
1018	VARIABLE DIMENSION WAS NOT A FORMAL PARAMETER	Can be caused by DIMENSION, COMMON, and all type statements
1019	MORE THAN 13 COMMON BLOCK NAMES	Compiles maximum exceeded.
1020	ILLEGAL FORMAT IN DECLARATIVE STATEMENT	Can be caused by DIMENSION, COMMON, and all type statements
1021	MORE THAN 3 DIMENSIONS	Can be caused by DIMENSION, COMMON, and all type statements
1022	TOO MANY (IN DECLARATIVE STATEMENT	Can be caused by DIMENSION, COMMON, and all type statements
1023	PARENTHESES DO NOT MATCH	Can be caused by DIMENSION, COMMON, and all type statements
1024	NO DIMENSION FOR DIMENSIONED VARIABLE	A variable in a DIMENSION statement must be dimensioned
1025	VARIABLE DIMENSIONED MORE THAN ONCE	Can be caused by DIMENSION, COMMON, and all type statements
1131	VARIABLE IN EQUIVALENCE RELATIONS HAS INCORRECT NUMBER OF SUBSCRIPTS	Can have only up to 3 subscripts
1132	ATTEMPTED TO RE-ORDER COMMON	Cannot re-order common by an equivalence relation
1133	MORE THAN ONE ELEMENT OF A SET IN COMMON	More than one variable in an equivalence set appears in a COMMON statement
1136	EQUIVALANCE RELATION ERROR	
1137	NO PROGRAM STORAGE ALLOWED IN BLOCK DATA SUBPROGRAM	Variables initialized in BLOCK DATA subprogram must be in labeled common
2001	ILLEGAL TRANSFER VALUE IN UNCONDITIONAL GO TO	
2002	ILLEGAL VARIABLE IN ASSIGNED GO TO	
2003	NO LIST OR ERROR IN LIST OF ASSIGNED GO TO	
2004	SYNTAX ERROR IN LABEL STRING OF COMPUTED GO TO	Illegal label value or missing commas separating label values (i.e., statement number references)

Appendix B—*Continued*

Code	Message Text	Further Explanation
2005	SYNTAX ERROR IN LABEL STRING OF COMPUTED GO TO	Illegal label value or missing comma separating label values
2006	SYNTAX ERROR FOLLOWING CLOSING PAREN OF LIST IN COMPUTED GO TO	Missing comma, illegal variable name, or extraneous information following
2007	ILLEGAL OR MISSING VARIABLE IN COMPUTED GO TO	
2008	MODE OF VARIABLE NOT INTEGER IN COMPUTED GO TO	
2009	SYNTAX ERROR IN ASSIGN STATEMENT	Unexpected operator, illegal constant, or illegal variable
2010	MODE OF VARIABLE IN ASSIGN STATEMENT IS NOT INTEGER	
2011	EXTRANEOUS INFORMATION ON RETURN	
2012	ILLEGAL CONSTANT ON PAUSE OR STOP	Constant not legal octal value or has more than 5 digits
2013	EXTRANEOUS INFORMATION ON CONTINUE	
2017	RETURN IN MAIN PROGRAM	Only permissible in a subprogram
3001	THERE IS NO PATH TO THIS STATEMENT, IT IS COMPILED BUT CANNOT BE EXECUTED	
3002	DECLARATIVE STATEMENT MUST PRECEDE DATA AND EXECUTABLE STATEMENTS AND STATEMENT FUNCTION DEFINITIONS	
3003	A LOGICAL IF EXPRESSION MAY NOT BE FOLLOWED BY ANOTHER LOGICAL IF EXPRESSION	
3004	A LOGICAL IF EXPRESSION MAY NOT BE FOLLOWED BY A DO, ENTRY, DATA, FORMAT, OR DECLARATIVE STATEMENT	
3006	LABEL DOUBLY DEFINED	Each statement number must be unique

Code	Message Text	Further Explanation
3007	LABEL ON THIS FORMAT HAS BEEN REFERENCED AS A BRANCH POINT	Cannot GO TO n if n is the label of a FORMAT statement
3008	THIS STATEMENT CONTAINS A BRANCH TO AN INNER LOOP OR IS REFERENCED AS A BRANCH POINT IN AN OUTER LOOP	DO ranges entered illegally at execution time
3009	LABEL ON THIS STATEMENT HAS BEEN REFERENCED AS A FORMAT LABEL	"Label" means statement number
3010	LABEL REFERENCED IN THIS STATEMENT GREATER THAN 99999	
3011	LABEL DEFINED ON FORMAT STATEMENT IS REFERENCED AS A BRANCH POINT	
3012	LABEL DEFINED ON AN EXECUTABLE STATEMENT IS REFERENCED AS A FORMAT STATEMENT LABEL	
3013	ILLEGAL NESTING OF DO LOOPS, AN OUTER LOOP TERMINATES BEFORE AN INNER LOOP	
3015	COMPILER CAN HANDLE ONLY 9999 REFERENCES TO VARIABLE DIMENSIONED ARRAYS IN THE I/O STATEMENTS OF A SINGLE PROGRAM UNIT	
3017	THE END LINE OF THIS PROGRAM UNIT OCCURS BEFORE ALL DO LOOPS HAVE BEEN TERMINATED	
3501	ILLEGAL STATEMENT REFERENCE IN DO SPECIFICATION	
3502	ILLEGAL CONTROL VARIABLE IN DO SPECIFICATION	
3503	ILLEGAL OR MISSING DELIMITER IN DO SPECIFICATION	
3504	ILLEGAL PARAMETER IN DO SPECIFICATION	
3509	ILLEGAL ELEMENT IN I/O LIST	"I/O" means input-output

Appendix B—*Continued*

Code	Message Text	Further Explanation
3512	ILLEGAL OR MISSING DELIMITER IN I/O LIST	
3513	UNMATCHED PARENTHESIS IN I/O	
3514	IMPLIED LOOP IN I/O LIST CONTAINS NO ELEMENT	
3515	SUBSCRIPT IN UNDIMENSIONED VARIABLE	
3516	ILLEGAL FORMAT REFERENCE	
3519	ILLEGAL SUBSCRIPT CONSTRUCTION	
3520	ILLEGAL TERMINAL STATEMENT NUMBER IN DO SPECIFICATION	
3521	CONTROL VARIABLE ALREADY IN USE	E.g., WRITE (6,10) (A(I,I), I=1,3), I=1,2
3522	CONTROL VARIABLE USED AS DO PARAMETER OF OUTER DO LOOP	E.g., DO 2 J=1,I DO 1 I=1,3 1 CONTINUE 2 CONTINUE
3523	PARAMETER MISSING IN DO SPECIFICATION	
3531	DELIMITER MISSING IN FORMAT STATEMENT	
3533	MULTIPLE DECIMAL POINTS IN FORMAT SPECIFICATION	
3534	DECIMAL POINT IN OTHER THAN F, E, G, OR D SPECIFICATION	
3535	GROUP COUNT PRECEDING LEFT PARENTHESIS SHOULD NOT BE ZERO	
3536	NESTING OF FORMAT GROUPS LIMITED TO 3	
3537	AN EXCESS OF RIGHT PARENTHESES IN FORMAT STATEMENT	
3538	A DELIMITER MUST FOLLOW EACH PARENTHESIZED FORMAT GROUP	

Appendix B—*Continued*

Code	Message Text	Further Explanation
3540	SYNTACTICAL ERROR IN FORMAT STATEMENT	Caused by incomplete format specifications, such as 10.2 instead of F10.2
3541	ZERO OR MISSING WIDTH SPECIFICATION IN FORMAT STATEMENT	E.g., F0.9 or F.9
3542	DECIMAL INDICATOR MISSING FROM F, E, G OR D FORMAT SPECIFICATION	E.g., F10
3543	DECIMAL INDICATOR EXCEEDS FIELD WIDTH IN F, E, G, OR D FORMAT	E.g., F10.11
3544	UNEXPECTED CHARACTER IN FORMAT SPECIFICATION	E.g., F10.H
3545	REPEAT COUNT OF ZERO IN FORMAT SPECIFICATION	E.g., 0F10.5
3546	ERROR IN HOLLERITH COUNT	
3551	IMPROPER OR MISSING DELIMITER IN DATA STATEMENT	
3552	NUMBER OF CONSTANTS IN DATA LIST DISAGREES WITH NUMBER OF VARIABLES	
3553	IMPROPER OR MISPOSITIONED ELEMENT IN DATA STATEMENT	
3554	ARRAY ELEMENT REFERENCED BY WRONG NUMBER OF SUBSCRIPTS	
3555	SUBSCRIPT ERROR IN DATA STATEMENT	
3556	FORMAL PARAMETERS MAY NOT APPEAR IN DATA STATEMENTS	Illegal use of dummy variable in a subprogram
3558	DATA STATEMENTS MAY NOT PRESET VARIABLES IN BLANK OR CHAPTER 2 COMMON	
3561	DATA REFERENCE IS OUTSIDE ARRAY RANGE	Reference to A(4) when A is dimensioned A(3), for example
3562	MULTIPLE DATA ENTRIES SHOULD BE SEPARATED BY COMMAS	
5001	ARRAY ELEMENT REFERENCE NEEDED	

Code	Message Text	Further Explanation
5002	SYNTAX ERROR-SUBSCRIPT NOT IN CORRECT FORMAT	General form of subscript is c*v+k where c = Integer constant k = Integer constant v = Integer variable (positive or negative)
5003	SUBSCRIPT NOT SIMPLE INTEGER VARIABLE	
5004	SUBSCRIPT NOT INTEGER CONSTANT	E.g., B(1.5) illegal if B is dimensioned
5005	NUMBER OF DIMENSIONS IN ARRAY REFERENCE AND ARRAY DEFINITION DO NOT AGREE	
5101	COMPLEX MODE NOT ALLOWED IN IF STATEMENT	
5103	LABEL FOR ARITHMETIC IF STATEMENT MUST BE CONSTANT	
5104	COMMA IS ONLY OPERATOR ALLOWED BETWEEN LABELS OF ARITHMETIC IF STATEMENT	
5105	CONSTANT IMMEDIATELY FOLLOWED BY (	
5106	) (NOT LEGAL - OPERATOR MAY POSSIBLY BE MISSING	
5107	ARGUMENT LIST CANNOT FOLLOW SUBROUTINE NAME WITHIN AN ARGUMENT LIST	
5108	NAME PREVIOUSLY USED AS A SIMPLE VARIABLE IS BEING USED AS A FUNCTION REFERENCE	
5109	OPERATOR PRECEDES)	Illegally formed expression
5110	TOO MANY LEFT PARENS IN STATEMENT	
5111	OPERATOR PRECEDES, IN EXPRESSION	
5112	CONSTANT OR VARIABLE CANNOT IMMEDIATELY PRECEDE .NOT.	Illegally formed logical expression

Appendix B—*Continued*

Code	Message Text	Further Explanation
5113	INCORRECT USAGE OF .NOT.	E.g.: .NOT.(A.OR.B).NOT.C
5114	OPERATOR OTHER THAN) TERMINATES THIS STATEMENT	
5115	UNMATCHED PARENS EXIST IN EXPRESSION	Erroneous parenthesization
5116	CONSECUTIVE OPERATORS IN EXPRESSION	Illegally formed expression
5117	.NOT. MUST BE FOLLOWED BY LOGICAL OPERAND	Illegally formed expression
5118	- MUST BE FOLLOWED BY ARITHMETIC OPERAND	Illegally formed expression
5120	ILLEGAL MODE USAGE OF LOGICAL DATA	Arithmetic mode is used with logical operator
5121	ILLEGAL MODE USAGE OF RELATIONAL EXPRESSION	Logical operand is used with relational operator; or integer mixed with real or double-precision modes using relational operator
5122	RELATIONAL OPERATORS MUST BE SEPARATED BY LOGICAL OPERATORS	
5123	ILLEGAL USE OF EQUAL SIGN	Multiple assignment statement not allowed; e.g., A = B = C + D
5124	EQUAL SIGN CAN ONLY APPEAR IN REPLACEMENT STATEMENT	Use .EQ. in logical IF instead
5125	ILLEGAL NAME ON LEFT SIDE OF REPLACEMENT STATEMENT	
5126	ILLEGAL ELEMENT PRECEDES + OR -	Illegally formed expression
5128	INTRINSIC FUNCTION NAME MUST BE FOLLOWED BY (	
5129	STATEMENT FUNCTION NAME MUST BE FOLLOWED BY (	
5130	OPERATOR MISSING IN EXPRESSION	
5133	ILLEGAL OPERATOR FOLLOWS ARRAY NAME	

Appendix B—*Continued*

Code	Message Text	Further Explanation
5134	NAME USED IN CALL STATEMENT PREVIOUSLY USED AS SIMPLE VARIABLE	Cannot CALL A if A is used as a variable
5135	ILLEGAL ELEMENT FOLLOWING SUBROUTINE NAME IN CALL STATEMENT	
5136	ARGUMENTS FOR STATEMENT FUNCTION DEFINITION MUST BE VARIABLES	
5137	DUPLICATION OF ARGUMENT NAMES IN STATEMENT FUNCTION DEFINITION	
5138	ARGUMENT IN STATEMENT FUNCTION NOT FOLLOWED BY , OR)	
5139	ELEMENT FOLLOWING) IN STATEMENT FUNCTION DEFINITIONS IS NOT =	
5140	ARRAY REFERENCES MAY NOT APPEAR IN STATEMENT FUNCTION DEFINITION	
5141	STATEMENT FUNCTION NAME USED AS A VARIABLE WITHIN ITS OWN DEFINITION	
5143	ILLEGAL MODE USAGE WITH EQUAL OPERATOR	Logical data is used with the equal sign
5144	ILLEGAL MODE USAGE WITH **	Logical data is used with the ** operator
5145	LEFT SIDE OF REPLACEMENT STATEMENT HAS UNDEFINED ARRAY NAME OR FUNCTION NAME	
5146	) FOLLOWED BY A CONSTANT OR VARIABLE NAME APPEARS IN STATEMENT	Illegally formed expression
5148	SYNTAX ERROR IN IF STATEMENT	
5149	CHAPTER 2 ARRAY NAME CANNOT BE PASSED AS PARAMETER	

Code	Message Text	Further Explanation
5151	INCORRECT COMMA USAGE	
9000	LABELED CONTINUATION CARD	
9001	LABEL IS ZERO OR CONTAINS A NON-NUMERIC CHARACTER	Improper statement number
9003	MORE THAN 19 CONTINUATION CARDS	
9004	FORMAT STATEMENT HAS NO LABEL	
9006	END OF STATEMENT OCCURS BEFORE HOLLERITH FIELD COUNT SATISFIED	
9007	MORE THAN EIGHT CHARACTERS IN IDENTIFIER	
9008	STATEMENT HAS CONTINUATION LINE(S) WITH NO INITIAL LINE	
9009	INTEGER VALUE TOO LARGE TO CONTAIN IN ONE WORD AND SINGLE PRECISION OPTION HAS BEEN TAKEN	Maximum $2^{23} - 1$, minimum -2^{23}
9010	UNIDENTIFIABLE USE OF PERIOD IN STATEMENT	
9013	INTEGER VALUE GREATER THAN CAN BE CONTAINED IN FORTY-SEVEN BITS	Maximum $2^{47} - 1$, minimum -2^{47}
9016	SIGNIFICANT DIGITS IN SINGLE PRECISION REAL REQUIRE MORE THAN 36 BITS	User constant given to more significant digits than can be retained in a real (floating point) number.
9017	SIGNIFICANT DIGITS IN DOUBLE PRECISION REAL REQUIRE MORE THAN 84 BITS	
9018	NAME ON END STATEMENT	
9019	NO RETURN STATEMENT IN SUBROUTINE	
9020	LABELED END STATEMENT	
9021	ILLEGAL DELIMITER	
9022	EXTRANEOUS INFORMATION IN COLS 1-5 OF CONTINUATION LINE	

INDEX